Endorsements for *Abandon E*

Adorno first wrote that to "write poetry after Auschwitz is barbaric", but then later that "perennial suffering has as much right to expression as the tortured have to scream". Singer's *Abandon Every Hope* cleaves these contradictory perspectives. And to cleave, as she tells us, means to split apart and bring together. This is writing that breaks and heaves with the bodies of animals whose infinite torture it presences. Through her facility with language, tone and rhythm, and an unmediated intimacy with her own brokenness in the face of inestimable suffering, Singer is able to bring the reader close to the violence and death that is foreclosed from sight and knowledge. This is a text that ought not and will never leave its readers.

DANIELLE CELERMAJER

A passionate and deeply thought-out work about one of our most troubling ethical blind spots—the killing of animals for meat. Singer's commonplace book, a blend of narrative, notes, literary history and elegy, unfolds during the Covid 19 pandemic—an event directly connected to intensive animal agriculture. Channelling a favourite saint, Aelwaer, who stands for rioters, quarellers and troublemakers, Singer's distinctive and courageous witnessing is a match for our century's 'special darkness'. She ventures where most prefer not to: into abattoirs where time is measured by 'the number of livers processed' and intensive farms where creatures become 'nutrient laboratories'. Between these hard truths she reminds us that animals are repositories of wonder. Witness the grief rituals of orcas, elephants and chimpanzees, the solace of our pets. If our mass killing of animals is a collective crime narrative, Singer's bold cultural detective work strives to understand its perpetrators and seek justice for the victims.

MIREILLE JUCHAU

Hayley Singer is a modern-day mermaid, a writer with extraordinary sensitivity to the junctions of human and animal life. This book does not shy from the depths—a glittering achievement on a profoundly murky subject. *Abandon Every Hope* is an otherworldly book of animal innerness and moral regard: a transformational guide to how we think about the creatures we consume.

From the booming heartbeat of a snake to prophecies wrought in chicken bones, *Abandon Every Hope* is a thrilling witchy book—deeply engaged with what it means to be animal in the early Anthropocene. A book of questions as cold and sharp as knives laid out in moonlight.

REBECCA GIGGS

# Abandon
# Every Hope

**Hayley Singer**

Hayley writes essays about literature and ecologies, queer embodiment and activism, multispecies in/justices and on reading and writing as worlds end and begin again. Her writing has been published in *Sydney Review of Books*, *The Lifted Brow*, *The Monthly*, *Cordite Poetry Review*, and *Writing from Below*. She teaches creative writing at the University of Melbourne. This is her first book.

Hayley Singer

# Abandon Every Hope

Essays for the dead

**UPSWELL**

First published in Australia in 2023
by Upswell Publishing
Perth, Western Australia
upswellpublishing.com

ISBN: 978-0-645-53699-7

A catalogue record for this
book is available from the
National Library of Australia

NATIONAL
LIBRARY
OF AUSTRALIA

Cover design by Chil3, Fremantle
Typeset in Foundry Origin by Lasertype

I did not know how to touch it it was all so raw

**Mary Ruefle, 'Kiss of the Sun'**

# February 2022

Something's smacking together in the darkened air where we are standing, on a darkened slope. Not a real slope, a slant of writing.

Here there is no time, only loads. Only the fact of shooting, only what we might hear in the writing: a smack in the face.

A smack so overwhelming, a face so overwhelmed by soft vertebrae crack. You could say it's just an ordinary smack. You could say they're just lazy swarms of gnats and horseflies. Just sheds, just piglets, just heads to be smacked against concrete.

Look at the face of a pig. Look long and slow. Look like you would absorb a beer. Look at this nauseated world of stale shed air. Look how each face looks back nervously with insects at the eyes, or blood, or

...

On a Monday morning when bodies restart egg-laying cycles. Look. Those chickens sagging at the knees, no hope they'll see the skies. Look. If I had to go mad—mind spitting along/teeth grinding—I'd start at the edge of an abyss like this on a Monday morning/oh god/do you hear that? A murmuring and

smeared with their own/oh god/and enter the warehouse of the liver guys early on a Tuesday, on a Friday, over and over again like sixty thousand times and more.

Have you ever seen anything quite so corpses as a fragment from a text that says *Four hundred livers later* as a unit of time? And says *that shit will fuck you up for real*? Though it is only a Thursday.

A Thursday. So common a day. So like every other morning of the greyest mornings. So daily in its grind. So daily in the smell of suffering souls, of inner organs split, of heat decay as global temperatures rise increasing *climatically stressful conditions* by hundreds of days per year by century's end.

                                          ...

Can anyone smell the suffering of souls? Of sadness? Of hell? Hell, I imagine, has a smell that bloats into infinity. Has a nasty sting of corpses ...

What was it Dante revealed? *Hell is a pit perfectly planned to contain confusion.*

How to write the smells of an egg factory? The question comes like a rustling in the paddocks, like silence, a long pause, like a Tuesday as I'm driving down the highway, like a moan when souls are dropped to the ground or hop lamely and foam, warm-pink trembling, sucking the edge of a wound.

...

You know that way you know, suspecting, on a Tuesday, that someone cannot endure the world anymore?

...

I drive past a smashed rabbit, a swamphen's wing, wombat broke open. Hear the sounds of smacking, but only in my head. Imagine picking up intestines. Should start packing a high-vis vest and traffic cones in the car so I can pull over and gently drag the dead off the road. Disposable gloves to check pouches. A shovel for roadside burials. Instead, I pull over and just imagine.

...

You know how language is a kind of food? How we commonly say that we butcher language, or else stuff words into each other's mouths? You know how questions of language can also be questions of life?

You know that way you hope everything weren't so damned sick? So stained? Like each face, teeth grinding? Like each truth that looks you in the face, grinding? Like each face above the neck held by hands wringing if they do not meet *the required standards*?

. . .

Abandon that hope now smashed against concrete. Pull down the warm blanket of *it does not really matter.* Pull it down, that blanket. You know the blanket I mean.

Like John Grant, the schoolteacher, when he gets drunk, high, goes roo shooting out in the fictional mining town, 'the Yabba', in Kenneth Cook's *Wake in Fright.* Goes *butchering this warm grey beast* and stumbling. Mind spasms. Roos limping, hopping, broken, while Grant sobs and rips another deep gash in the furry panic of the night: *the blade went easily, deeply, but the kangaroo would not die.*

It's a nighty-night blanket of *it hardly matters* that comes over him as he drives the knife. Everyone back at the car laughing. Grant covered in blood. They wash it all down with beer as they cut roos to pieces even before they're dead. Then dying, finally, in answer to the demands of too many incisions. And that warm drunk blanket numbs the mind against the slicing and smacking of heads, tails, hooves of hearts hammering in all their multitudes, of flanks smeared in the coffin of other lives, of bodies of dirt on all fours, of bodies of pus or blood. Bodies pushed into the teeth of their last ever morning.

Imagine your last ever morning. What do you hear? Silence? A moan, a wail, a voice? Nervous voice of a poet, smeared with insects at her eyes, who says *Let us descend* and so you descend.

# To Be Struck

In 'What is Contemporary', Italian philosopher Giorgio Agamben asks what it means to look into the *special darkness* of one's time. Your eyes, he says, will be struck by beams of darkness.

Say you let yourself get taken by this darkness.

Say you get struck by beams of such special gloom and drown, like Yoineh Meir in Isaac Bashevis Singer's story 'The Slaughterer', a man who might have become the rabbi of a small town and is instead appointed ritual slaughterer, who does not want to kill but day after day must fulfil his role slaughtering hens and roosters and geese and ducks and cows and goats. With every tremor of the bodies he holds in his hands he feels a tremor in his own flesh.

Slaughtering, he decides, though ordained by God, is a punishment that has been laid upon him.

After months of killing, he longs to escape the material world. But his house is full of meat. The scent won't leave his nostrils. Air and bones and water all carry it. His days become nightmares. His nights fill with impossible animal returns.

Meir feels sick, sad, guilty, insane. Every slit gullet and every plucked feather congeals in his soul marking his humanity as inhumanity.

*The whole world is a slaughterhouse*, he cries and days later is found floating in the river.

A key factor of contemporariness, writes Agamben, is that those who are truly of their time cannot coincide with dominant trends or ethics.

Earlier in the century Gertrude Stein said, *For a very long time, everyone refuses and then almost without a pause almost everyone accepts. In the story of the refused in the arts and literature the rapidity of the change is always startling.*

How is it that we all come to decide? Are there steps to this cycle of acceptance, or is everyone suddenly struck by magic?

The creator of a new composition is an *outlaw*, off the grid, aesthetically, ethically and politically unspooled till their vision is safely rendered past and the work becomes (if anything) a classic. Such dys-chronicity is the shared terrain of the contemporary, the writer and the ghost.

*A spectre is always revenant*, says French philosopher Jacques Derrida. *One cannot control its comings and goings because it begins by coming back.*

Ghosts flesh matters thought not to matter. Writers share this in common with ghosts.

The task, the urgent task, is to figure out which beams of darkness any of us should be staring into right now.

# On Immunity

Today, pictures pass through my head:

> Shots of frontend loaders clearing bodies off beaches.

> Shots of three-hundred-or-more bodies washed from round Murwillumbah to the beaches of my childhood.

> Shots of bodies wrapped in sea foam. Of cows jammed into trees or grabbed by tides, clinging onto dinghies. *Days of gore.*

I want to fall into a community of grief where only things that can be done in grief—singing hymns, pouring water into bowls, covering mirrors with thick cloth, building altars—are done.

I've got no capacity to attend to the actual—dishes, driving, meetings, showers. I'm slipping from moment to moment back to the beach where a brown cow washed up, alive, after tumbling for hours in flood waters. After floating up the Tweed River. After losing her baby to the waters, she/the brown cow/ came up wandering, and ran, and was eventually shot five times by police 'cause she wouldn't let herself get caught.

That happened in the north. I'm in the south reading, watching, feeling uncoordinated as an outsider with dry feet.

I barrel up the highway, past other cows loaded on trucks, sniffing the air. Try to imagine a metaphor or simile for them. Nothing else under the skies quite like those trucks: twenty-first-century shepherds of outright annihilation.

Jerky now at roundabouts, driving in snatchy movements missing other cars, but only just. Eyes glazed. Yawning and nervous thinking, asking no one: How can I speak the mutilated world without mutilating it further, or again?

Keeping a single glazed eye on the cows in the truck. As it turns, I take a wild turn, too. So now I'm following it. Watching the cows taking in the breath of the world. Air full of their ancestors' gaseous ghosts. Barrelling across the skies. Fugitive astral globes.

Just the other day, a woman looked into a paddock of cows and said, airily, *they're happy*. Though they were screaming their full-to-bursting udders and someone had to call the council because the farm didn't have the capacity to milk

and when I say something about multispecies PTSD I get that sideways look, even as two rams' heads are found up in the top paddock, even when a man wrapped one of the heads in plastic and kicked it, football-like, to the front door of the house surrounded by tall grasses now. The rams were, after all, supposed to be the lawnmowers.

There's a collective animal trauma against which there is a collective human barricade. We cannot *actually see* sadness and pain or hear a cry of heartache when we see sadness, pain or hear a cry of heartache.

I'd seen a lamb, tangled in rope, tied to a post and tried to approach till dogs came barking and a woman emerged from the house.

I'd heard gun shots in the afternoons and evenings that time Woolworths, somewhere in South Australia, went viral for writing in their meat aisle *What do you call a cow with no legs? Ground beef!* A joke that twists laughter with a knife, that gives immunity to 'the human' from the life of 'the animal'.

Narrative is used like that, like a vaccine to give 'the human' what is denied other animals.

I gather and carry necropolitical narratives—that subjugate life to the power of death—around my neck as remembrance items. Thinking, how am I to re-narrativise and extend those stories? How to re-form the flesh of the familiar? To make manifest the deadly unfreedoms that shape the world of minutely organised and profit-driven death?

The cultural habitat of the meat aisle is dominated by conventions that decide what can and cannot be shown, what is and isn't sayable. Meat-speak is a highly regulated but spectrally contaminated genre, thought to be no genre at all. It operates endlessly in the background of everyday life, shaping cultural permissions and prohibitions of bodies that can be eaten and bodies that eat.

This makes me think of narrative as war by other means, which is a rephrasing of Foucault's understanding of *politics as war by other means*, which is to say that narrative is one of the theatres in which the *war on animals* is fought.

A critique of narrative might begin with a critique of the unacknowledged ghosts of politics, philosophy and story. In

*The War Against Animals*, Dinesh Wadiwel conjures those who have been skinned to make Marx's leather boots and who haunt the boots. So, too, a critique of narrative might begin with a critique of 'the human' and an opening up of narrative to the ghosts sliced out of bodies cut to pieces.

Back on the road the truck turns into the livestock exchange near Pakenham. I'm carried onto the highway, and on and on to work thinking of how to rephrase and reposition stories about the lives of animals. Us animals. Thinking of what the cyborg laureate Jillian Weise has written: that scholars who venture into the world of animal studies can *rest assured that their subjects remain silent*. But if only that weren't so.

# August 2021

Bataille knew it: *the slaughterhouse is cursed and quarantined like a plague-ridden ship.* This thought is fixed in my head. Fixed is the idea that the world of the slaughterhouse produces and ignores slaughterhouse novels. Fixed, too, is Jean-Baptiste del Amo's novel *Animalia*. Just as the vision of knives sharpened on the barrel of a gun thrust into a nightmare called a piggery is floating in my mind. All leave bloody handprints on my eyes.

It is not just that a hand must grip a knife, but that a knife becomes a hand, becomes a way to touch, which is to kill, which is the world of del Amo's novel where summer has come and an epidemic has broken out in the pig sheds.

The pigs have been plunged into an even more putrid world than before. A man works alone tossing stillborns and placentas into buckets for the incinerator. He is trying to save the family piggery. The smell of slurry is up his nose and burning.

The shit is stupefying. The shit is killing him. It is the shit that advances, not the man. The shit that becomes more impenetrable as he progresses. It is the shit that senses his motions, not the other way around. It is the shit of his rummaging or in which he stands—now moving, now motionless. The shit says *you've got a tragedy on your hands.*

Shit records that the world is being unmade. That the ground of all being is being unmade.

What can he do? Take things in hand? Lose track of time? In its proliferation this shit is paralysing and in it one finds what is just and what is unjust. The shit does not negotiate. It flows. He believes those who stand gut deep in it are bodies and no more than bodies, which is a thought that wrecks the life of the word *life.* In the shit he finds eternity, only not. Time is indistinguishable from shit in these sheds. In its magnitude, this shit has become a weapon.

The pigs who live in the shit are mostly invisible to everyone, even him. Their prolonged and agonised cries convey only a limited dimension of their suffering. Their wounded bodies bear the unbearable. To live as tools. To be broken in stages, through severe physical restriction. Tied to a post. Clamped to the floor. By disease or overuse.

In the sheds, the labour of bodies meets the labour of war, which is the endless transformation of the world into a killing machine. Or, the transformation of a series of banal-seeming technologies into deadly constellations. Like, truck-fence-knife. Like, ship-heat-shit. Like, bars-hunger-boredom.

The pig units are two metres by three and each contain five to seven pigs. The pigs cannot wallow in mud. Cannot express pig culture. Cannot regulate their body temperature so the *pig units become like furnaces.*

Their work, pig-work, is the work of surviving physical pain, mental frustration, boredom, anxiety, being convulsed by spasms. This work, this unseen work, unmakes them. Their bellies, their skulls half-eaten with boredom, pain, heat, heartbreak

as they are unmade they make the raw materials of capitalism: muscle, skin, bone, blood, force, physical force, and other bodies that will be similarly broken down in stages. And now it seems that every second of the man's life has been leading to the moment when a blinded pig follows as he weaves between pig units, while another runs screaming for the boundary fence. She hits a slurry pit. Fall ins. Drowns.

Soon the man in the shed spends all day handling genitals and skin and treating purulent metritis, which is the inflammation of the uterus announced by a watery red-brown discharge. When the pigs shriek, he shrieks too. He can no longer leave the sheds, the pigs need constant tending

he sleeps in the adjoining office. Mutters to himself. Comes to shit and piss on the concrete with the pigs. But still, he knows himself to be a man, not a pig. How? Like he has a natural immunity. Like he cannot carry the particular bacteria found in these pigs so he is a man who screams as he shovels shit, who slips into the land of the dead each night. And wakes to the land of the dead each morning.

## May 2020

*We fell asleep in one world and woke up in another,* wrote the poet Haroon Rashid. In this new world it is winter. A plague has broken out here, too. *Some believe it to be the child of a bat.* It had *re-invented slowness.*

In this new world a man shoots six grey-headed flying foxes by the side of a road not far from where I live. One dies with half-chewed fruit in her mouth. Another guy beats flying foxes to death in his backyard; a mother and daughter are accidental witness. All over the world people go on bat killing sprees. Mobs descend on caves with lit torches

onions disappear and garlic I take photos of empty shelves. Send them to no one. Students thin out. Reappear. I teach. Go home. Journey back again. Dream of buying a huge box of Roma Truss and Beefsteak tomatoes, of taking it home and calling it My Red Fate. But they, too, are gone.

. . .

So I'm re-reading a story called 'A Chicken' (Uma galinha) by Brazilian novelist Clarice Lispector. It goes: one Sunday a family is about to destroy a chicken, 'cause she's *merely* a chicken, 'cause it *wasn't yet nine in the morning*. But the chicken has other ideas. She leaps to a neighbour's terrace. Flies free. Navigates alien topography. Runs across buildings. Fear flaps all over the skies. The Father in hot pursuit finally brings her back to the kitchen. So overwhelmed she opens her wings and lays an egg. Becomes profound to the girl who watches dinner bring new life into the world. The girl cries out to her family. They come see: the hen, the egg. Look, *look!* The egg, from the hen! They vow never to eat chicken again if this hen is killed.

Turn the page and you'll see they kill her, eat her, the years go by. What luminous betrayal.

...

I sit beside myself on the couch. Time alone touches me. Dark outside. I stare at my keyboard. What a sad year each day has become. If someone asked what today was like I'd say: *Imagine being handled by a giant.*

Then it is a Tuesday, to my great surprise. I put on clothes and walk to the garden shed, which is my studio. I spend three days on this book. I work badly, then teach, then try to smile all day because smiling all day mounts an attack on normal consciousness. I wander through books and call this time *my mental commute.* Then, simply, or finally, it is spring for a day.

. . .

Sleepless nights smack me sideways. It is also a time of strange dreams. In one, a Subcommittee from the Council of Poets debates whether the primary guide for imagining and describing the contemporary Earth Era should be faecal matter. Once agreed upon, the official substance will guide all subsequent poetic analysis of Earth. Will be used to describe the present as a *faecological time interval*, in which many conditions and processes on Earth are understood to have been profoundly altered by shit.

. . .

The American writer Joy Williams has this story about God, which is also and mostly about shit. It is also about the grace and intelligence of pigs. In it, children are taken on a school trip to a slaughterhouse. The logic being that kids need to know what real work (as opposed to intellectual labour) looks like.

The kids are bussed to the facility. Turns out they can't go in. Don't see or hear a single pig. Instead, they see a vast brown lagoon of waste. They smell a smell they have never smelled before.

Slurries / open air cesspools / manure pits / anaerobic lagoons. Whatever you want to call them, their job is to store faeces, urine, blood and 'other' fluids, like meconium and miscarried piglets, and the smell that comes off them has its own weather.

When a pool reaches capacity, the contents might be liquified. Sprayed onto nearby fields. Undigested antibiotics fed to the pigs are sprayed across the fields, too. Communities who live nearby such operations speak of the way the stench burns their eyes, their lungs. So matters of air are thickened with piles/loads/hunks and other quantities and configurations of shit.

Unlike other gases that might bubble off, hydrogen sulphide sticks to the manure molecule and accumulates with the build-up of shit. It is heavier than oxygen. A colourless gas. In low concentrations it smells like eggs. Unpleasant, but not deadly.

Once it reaches higher concentrations it paralyses your sense of smell. Can't be detected by the human nose. People have been lulled into a false sense of safety by this.

I remember reading, this is years ago, about a summer afternoon in Ohio when a farm worker entered a ten-foot-deep manure pit to repair a pump. Overcome by the smell of the pit he passed out, fell into the pool, drowned. His fifteen-year-old nephew tore into the pit, attempting a rescue. He, too, was overcome. The boy's father ran in after him. His cousin. His grandfather. One after the other. Like some sick joke. All five members of the family hit the bottom and drowned, which strikes me as a trick of industry: when the outside reaches in and the air no longer breathes you.

The gas enters the body through the lungs. Dissolves in blood. Makes its way round. The gas's primary itinerary is death. In low chronic concentrations it causes headaches, low blood pressure, nausea, chronic coughing, depression, tension, anger, fatigue.

Air, turbulent participant in creation and destruction of worlds. Your breath can be seized or delayed by beats. Have you read the poet Anne Boyer, who has written that *Only villains attempt to invert the lightest, purest thing—air—so that it becomes the most poisonous and the heaviest?*

# Back-alley Poets

Who has ever really met a voiceless animal? Or is it that they are, as Walter Benjamin wrote, the *receptacles of the forgotten*? Forced to live as if their tongues have been cut out until their tongues *are* cut out. This makes me want to paraphrase Sophokles' suffocating idea about women and silence: silence is the *kosmos* of animals.

...

*It is in large part according to the sounds people make that we judge them sane or insane, male or female, good, evil, trustworthy, depressive, marriageable, moribund, likely or unlikely to make war on us little better than animals, inspired by God.* This is the opening of Anne Carson's essay 'The Gender of Sound', where Carson scrutinises the history of ideas that have shaped voice, vocal quality and vocal use as part of the ground on which 'the human' stands.

A very specific kind of vocalisation makes 'the human,' writes Carson. Specifically, *manly* pitches, speed. Specific relations to ideas of coherence, grammatical orderliness and self-control secure a complex immunisation against animality.

Elaborating on this zoological doctrine of sound, Carson notes that Gertrude Stein's biographer made a point of saying that Stein *laughed like a beefsteak* and loved to eat beef. By this mess of figurative and factual moves, Stein is walked off the ramp of humanity and into animality, made to stand at a dividing line, as if she were a piece of unfinished business.

...

In fairytales, snakes talk. Crickets, doves, wolves, rabbits, mice all talk. In the Bible a donkey talks. Orpheus was considered exceptional because he could talk to other animals in *their* language.

In fairytales, animals are talking even when they're dead. Sabrina Orah Mark calls animals in fairytales *feral poets*. Like they are back-alley poets, edge of fire poets, chasing car poets, bin-diving poets, borderland poets. Not wild not tame not domesticated but syanthropic.

*Syn + anthro:* together with + man. Animals who live in houses, on farms, along roadsides, in rubbish dumps.

A syanthropic poet: the poet who exists in a world that is domesticated, yet uncontrollable. The poet without habitat, yet lives. She is an expansive, opportunistic and emergent poet who feathers her nest with cigarette butts.

I am not trying to start a myth of the poet as down-and-out-animal prophet. I am not suckling romantic notions, just trying to locate my animal poet body in time and space. This poet will never have it as hard as any other animal on Earth.

...

Whenever I listen to Stein read her work, I think hers is a voice that could tear open the sky. A roar and a boom as worlds begin and end.

Stein's lover, Alice B. Toklas, wrote that crime is the gateway to cooking, that eating is pleasant but murder isn't. Murder is what makes cooking truly terrible.

First came war, then the Occupation. Toklas learned to shop at a restricted market. To buy and cook what was available. One day during the Occupation, a fisherman didn't have the time to kill a carp, so Toklas severed the spine at home. The moment she'd made the cut, she staggered back, fell into her chair with bloodied hands, lit a cigarette and waited for the police.

# Radiance of Nothing

I sit late into the night researching cow abortions online for a thing I am trying to write. I take a virtual tour through the circuits and histories of bovine parturition trauma as it is displayed in an expansive online archive that throws liberationist videos together with footage of bobby calves getting shot in the head soon after birth.

Then I sit in the dark and all I can think of is nothing: the ebb of nothing, radiance of nothing, mutilated life of nothing.

Nothing has real form and I think this formlessness has taken up residence in my body.

In the morning I write notes to myself like:

> Ask strange questions. Track the compatibility between violence and rationality. Search for meanings that are not self-evident. Trail the creation, development and ongoing shifts of toxic cultures, GMO, cultures of necropolitics and of worldwide dispossessions.

> Resist the seductions of narrative coherence. Pitch your tent with those who reach towards explosive, inventive and

creative lines of more-than-human communication, and who are often called hysterical, mad or irrational.

Challenge ideas of human uniqueness. Write along multiple and overlapping axes of difference including becoming-animal -vegetal -toxic -insect -meat. Write along the lines of grief and by doing so allow grief to matter.

Take note of how you are becoming sick from this world.

Take note of how you are becoming sick from lack of it.

Track how things change: the fast and the slow, all that melts and what solidifies, too. Of how narrow things have become, like the way cash has become synonymous with life. Do not write this in a generalist or partial way. Take interest in what you do for cash. Take interest in your genealogies and their historic relations to cash, and by extension their historic relations to certain kinds of life and death. Do this in the aesthetic mode of poetic production so it becomes a new kind of knowledge.

Track the ways your work and poetry change over time. Track the stealth politics of your poetics and how these change over time. Track how the poetics of your politics disrupts your access to cash, which is to say disrupts your access to a certain kind of life.

Of life which is movement, sensitivity, respiration, nutrition, weeping, dancing, singing and also barely even existing at all, and which translates to a poetry of organs and roots, collective and individual.

Write your own definition of this thing called Life, which is to call or name a thing you have made or that you seek to unmake. If the latter, skip the step of naming and go on unmaking. Give yourself a moment to look see what you've made through the process of destruction.

Take time with this thing you've unmade, like the way you observe the collapse of weather patterns. Step outside. Look at the sky. Take your jumper off. Check all your apps. Try not to make perfunctory observations, or in the manner of remote observation, or in the absence of evidence. If there is no water or electricity, if there is too much heat, too many sirens or not enough, say that there is no water or electricity, that there is too much heat, too many sirens or not enough.

# Manifesto

A pink thighbone sticks out from a piece of roasted chicken. The colour is carmine, actually, which is made from the pulverised bodies of female scale insects *Dactylopius coccus*. The insects are used in cosmetics and beverages. It takes seventy thousand of them to make less than half a kilo of colour. This carmine bone has been made for The Pink Chicken Project, which proposes to change the colour of the entire species of domesticated chicken *Gallus gallus Domesticus* so that the pigment can be fossilised in the Earth's rocky strata.

The investigators imagine visually reoccupying the current geological era. Offer future geologists a poetic marker for humanity's devastating effects on Earth. Using a newly invented gene editing tool, the investigators propose encoding a warning against ecological damage in each chicken's DNA. They say the warning should read:

> *We, the humans of planet earth, write this message at the beginning of the Anthropocene.*
>
> *The current devastation of the planet is not the result of activities undertaken by the whole species* Homo sapiens: *instead it derives from a small group of humans in power,*

*upheld by the injustices of white supremacy, colonialism, patriarchy, heterosexism and ableism. We urge you to fight this oppression: for it enables and aggravates the anthropocentric violence forced upon the non-human world.*

*Sent in hope that you have re-imagined us as a biological organism ...*

Advice I've been given about writing manifestos:

1) Keep it short—only white supremacist terrorists write long ones.

2) Expect no one to follow it.

# Abandonment

Though predicated on the production of mass death, which is a form of mass loss, industrial slaughter cannot be described in the language of loss. Those born into the system are endlessly replaced, endlessly replaceable, and can only be written in the language of abandonment, which is the flip side to loss, or to getting lost, which speaks the shadow potential of being found, rescued, traceable, countable.

To be lost is to walk through doorways into an unknown, into the transformative. It is a darkness. An important darkness that holds gifts. This is the species of lostness Rebecca Solnit writes about in *A Field Guide to Getting Lost*. This is a book I find magnetic because it is about distances and desires that are blue and spiritual in an earthbound away, because the earth is spiritual and blue in its distances and desires.

Years ago I travelled to Chicago to take up a fellowship at the University of Illinois. Either I did not take the journey from the hotel L and I stayed at in River North—built exactly one hundred years after the first slaughterhouse was constructed on the north branch of the Chicago River—to the site of the old meatpacking district, The Union Stock Yard & Transport Co. Or, I have forgotten that particular part of our time in

Chicago, a city of glass and steel and art and jazz bars and broken sidewalks. Chicago. Ruled by a strict, visible order of grids, which feels endlessly reproducible, which feels militant and formulaic, and in which I was constantly getting lost.

*A Field Guide to Getting Lost* is mostly about losing one's way or oneself, entirely, as finding a new way or a new self, entirely. The book slides through the world towards the limits of knowledge, and sometimes beyond. As Solnit writes, in lostness you collaborate with mystery or the unbounded unknown. And if lostness is an orientation that points towards life beyond the bounds of the known, it is directly opposite to arrest and confinement, which is the cardinal point of abandonment, which is what I am hung up on.

I carry facts about the stockyards that I learnt at the time— that they were closed rapidly after World War II, that much of the original architecture of the place, including all the wooden pens that held so many millions of animals, has been dismantled—but any embodied or sensory experience of the place has vanished from my mind.

The train journey from Union Station is what I remember.

The rural college town, Urbana-Champaign, where I would stay in a cold dorm room, feeling lonely, on the agricultural campus of the university, is what I remember.

I was there to think about animals, write about animals, talk with others who were thinking and writing about animals, too: horses, whales, rabbits, dogs, cows.

I met a man there who was an actor and nightly turned into a bear. He led audiences through a forest that once provided shelter and food for his bear ancestors but could do so no longer.

I heard the slow, tidal heartbeat of a snake projected into a room through a stethoscope connected to an amplifier. The woman with the snake wanted to share something intimate between us, the writers, and the snake because people fear and even hate snakes even though they are so likeable. Some of them live in tight-knit communities, she said, they hang out with friends, she said, they care for their young and the children of their friends, too.

Some nights, when I wasn't writing, I laid out on the open grass of the university lawns drinking wine, watching fireflies in their mating-lighting display, though at the time I didn't realise I was inserting myself into such an intimacy.

...

As I left the city, I took a photograph of the south branch of the Chicago River through the window of the train as it paused on the Canal Street railroad bridge, which is the only one of its kind to cross the river.

The train slid over the river away from the glass, iron, steel, terracotta and raw concrete skyscrapers, which grew small and hazy and blue and, finally, becoming so small and hazy and blue appeared like nothing more than air and light when they shimmer and bend in allusive collaboration between hot earth and cool air. Broad palms of the ground opened up, and so did the sky. First, it was just a paste of blue. Then there was enough blue that I saw how the sky was an entirely new world. I fell asleep and dreamed loud, wild dreams.

When I woke, I tried to envision the prairie grass that once occupied the prairie peninsula, that once occupied the place of the city and far beyond to the South, West and Northwest, and through which I was travelling, that had been so badly damaged, if not wiped out by agriculture and construction in the late nineteenth century. I imagined the grasses once rolled in the wind. More ocean than land.

I thought about the city but failed to unsee it, to peel back the concrete. So I gave the towers names of prairie wildflowers and grasses—swamp milkweed, purple prairie clover, switchgrass, hairy wild petunia, big bluestem and compass plant—and I tried to picture those skyscrapers consciously taking on other forms like whale skeletons, water lilies, DNA spirals, a bird's nest, limestone outcroppings, sunflower petals or a giant eye. Only these were failed emulations of flowers, grasses, sedges and rushes. Only they weren't failures at all since the point of this city was the transfiguration, not the imitation, of nature.

...

The logic that blurs the boundaries between what is natural and technological is embodied in a project, completed in 1900, that reversed the flow of the Chicago River so it no longer drained into Lake Michigan but moved away, to empty into the Illinois River that flows into the Mississippi. The point was to take the city's sewage —including the offal, some bodies, or body parts, and manure from the stockyards—away to keep the lake, the city's source of drinking water, clean.

At this time, pigs and cows were being pulled into a similarly strange hybridity of nature and technology. More specifically, of machine and organism. Their bodies burdened and changed by commodity market chains, new and rapidly changing transport routes, feed and slaughter routines that were governed by new tax systems, credit systems, new kinds of crops, climates and politics.

And all this would, in time, almost entirely uncouple them from the world of natural things. Which is exactly where we—I mean they, those animals, or, more precisely, their bio-homogenised ancestors—are now.

The first commercial meatpacker was set up in Cincinnati, Ohio Valley, in 1818. Workers packed pigs into a large pen next to the packing plants and would walk over the back of the pigs and strike each one on the head with a two-pointed hammer. The development of a rail transport system—that allowed animals to be moved live from east to west, all handled through Chicago, while others walked from prairies to places of slaughter—as well as blockades during the Civil War, meant that the old 'Porkopolis' of Cincinnati was overtaken by Chicago, which was placed to become the food depot of the Nation.

At night, on the lawn, sipping wine, I read that in the killing season of 1863, the slaughterhouses of Chicago eviscerated 970,000 hogs and by the next year the Chicago Pork Packers Association moved to establish the Union Stock Yard which would be accessible to all—drovers, stock-raisers, brokers, commission-merchants and buyers. At its height, the Stock Yard would turn more than eighteen million animals a year into meat, glue, margarine, gelatine, leather and fertiliser.

The site of the stockyards was marsh swamp that had been drained. Plank walkways, brick buildings and wooden sheds were constructed on the site prior to its opening on Christmas Day 1865. And now it's an industrial park.

There's a photograph held at the Art Institute of Chicago that shows a Hereford steer standing at the gateway of the stockyards. He stands beside two men in white suits. They're smiling and shaking hands. Above them hangs an enormous sign welcoming this steer, who is the billionth animal to pass through the gates. He stares past the frame of the photograph.

The steer's name is Billy, he came from Iowa and was sold for one dollar a pound. Billy the Billionth, that's what he was called in *The New York Times*. Beyond that, the details of his life are lost.

Getting lost is not the opposite of death. Though it does involve risking death. While the details of Billy's life are lost, Billy himself was abandoned. To be abandoned is to guarantee death in unfathomable ways. To be lost is not at all the same as being abandoned. To be lost can mean someone desires you found again. Abandonment is a place bereft of this desire. To live through abandonment is to live in the mode of disappearance.

Billy was born to disappear, and in his eyes I see the look of one who has been abandoned. His eyes aren't despairing but dazed, or stupefied, or it's as if he is staring into the future, seeing his own violent destiny play out, or like he's walking through a dark tunnel straining to see what's up ahead, or he's caught himself in that moment, realising he is midway through a journey to some disastrous unknown.

. . .

Those who have been abandoned are not irretrievable, though mostly it feels that way. The language of abandonment is the same as the language of being buried alive. You are lowered into a hole. Covered with dirt. As in a horrific dream, you know in that moment you are not recoverable. Neither by knowledge, nor by memory. If you live, you live forgotten, with dirt packed firmly into your mouth. Which is impossible. Which is fatal. Which is the point.

The point being not to breathe better or shake off the dirt that reminds you who you are, who other people think you are. The point is to disappear even, and perhaps especially, when you're in plain sight.

# Big Curse Energy

What place is there for poetry in a world of ongoing and emerging zoonotic disease? A place where life is pulled from life as fast as raging rivers rage?

I move through the day hanging off the edge of these questions. In class, I suggest we all put our heads down and write. The strange sounds of student fingers on keyboards make a huge impression on me.

There are riots. An earthquake. Every moment is a car shooting off a cliff. I look into my computer. It is flickering and shaking. Or I am flickering and shaking. Turns out I need as much wine as I can get.

In class we agree to dive under tables if necessary. L watches footage of angry men assaulting the traffic on the Westgate Bridge. *Why Horses? Why are they singing Horses?* We finish a bottle. Open another. We try to recall the lyrics.

Now stickers are stuck around the city blaming Jews for 9/11. A rash of swastikas. Electrical storms of racism, xenophobia, gather and break. Like giant capacitors, hate roams the skies. They strike and kill we know such storms can develop in any geographical location. Blood pools on street corners. Oceans fill with bones.

I stir a big pot on our small stove. It's filled with water and onions and salt. Into it I throw words like 'alien', 'parasite', 'degenerate', 'blood and honour' (Proud Boys), 'blood and soil' (Nazis) as if they were sage, pepper berries, fingers, eyes, tongues. I stir and stir, watching it all bubble and burn

it takes hours. Then it's far too late for dinner. *Dinner's poisoned*, I say a little too loudly. I've had too many gins. I drink when I want to. I drink when I don't.

I wobble out to the very back corner of our garden and pour the hell broth out. I should say something, I think. Some magic words. And google, *How to break a curse?*

I lie awake turning over the dried bones of the day.

I draft long overdue emails in my head to students, colleagues, editors, managers, collaborators. *I'm sorry, I'm sorry, I'm sorry for my crocodile!* is how each begins and ends.

I walk and walk in bed, almost to the point of sleep.

I walk beyond sleep. Beyond patience. Patience has been shaken from me like dried seeds.

I tuck my head's comings and goings under my pillow. I breathe. I am breath itself. I am lavender. I am fading light. I try to be this and other things all the way till morning when I am tearing at my cheeks. Eyes cornflake crunchy.

I turn on my computer, it wants me to demonstrate I am worthy of it but I resist.

I am actually a boat, caught at the top of a wave. I spend a great deal of time labelling my emotions with the names of pastoral paintings: *Landscape with a Calm, A Peaceful Nibble, Landscape with a Storm, Nothing New Under the Sun, Landscape to Swallow You Whole, Spring Blues.*

Finally, students and I gather online. Each one of us like a princess in a tower, in a certain kingdom that is neither really here nor elsewhere. One by one we carry forth our questions. The first is a river we must cross. The second a forest. The third turns into a story about a wife who cuts the head off her husband's favourite horse with a sabre.

There are public tantrums. There are riot police. Crack downs. We watch all of it on YouTube. We drink too much. We do other quiet things in the extreme. These are slow changes that go unspoken in our private world: sudden heart pains, smouldering aloneness, losing friends, thinking about what you could be doing differently, doing nothing differently.

I get up and wash my face. Pick a knife out of the knife block. Immediately put it back. Crack some pepper into my left hand. Walk outside. Anoint the garden with this pepper. A breeze comes in a blur. A silvery curve. A watery slap.

# Never Destroyed

Once, a poet invited other poets to crowd into a seminar room. The lights were dim. The poet leaned in, *Have you ever thought of writing a poem that could never be destroyed?* he asked.

*Is this a scheme? Am I in a scheme now?* are the questions I did not ask. But no. The poet just wanted to be the first to have a poem encoded on a new synthetic genome to be written into the DNA of 'meat' chickens. *Just.*

First, he would write the poem in English. Then it would be translated into DNA base pairs and those base pairs would act as a watermark on the new synthetic gene.

Afterwards, I go downstairs. Everything is better outside. Outside, I can breathe. And I wonder if I was really awake to the presentation. Or if I was stepping into another country of words, which may, or may not, have been shaped around the novel *Moby Dick*

*Let faith oust fact; let fancy oust memory; I look deep down and do believe.*

I have to tear down a wall to get back to this memory and ask it questions. It is an audacious wall against remembering. A straight wall. Tall. Which is to say, I have to sabotage something of my mind to remember this moment correctly. I have to learn how to swing a hammer straight through re-constructed memory stumble around in the real past.

Of all the birds on Earth perhaps only *Gallus gallus Domesticus* will survive our time—this raging little crush of history. I mean survival as a species not as individuals with the ability to outlive, transform, recover or heal after trauma.

I mean what Shakespeare wrote: *How many goodly creatures are there here! ... O brave new world.*

## May 2021

The world is too full and too empty, I say, sitting still. Actually, I am saying nothing. My silence startles me. The problem is that my mouth keeps cutting out like a flooded engine. The problem is that there is an invisible force, a wave, hitting down on me. Big watery slaps that threaten to split me in half as if I were a piece of driftwood. I need to tell myself a story about reality.

...

Post-mortem, wild bird bodies are decay and scavenger prone. Domesticated birds sold for consumption and discarded as drumsticks, wings and whole-body carcasses are dumped in landfill waste sites, they end up in places with low anaerobic activity. Rather than disintegrating they are mummified like fictional monsters; contemporary broiler chickens have supersized skeletons. Their genes altered so they are always hungry. From one day old, they live under electric lights to stimulate their eating. A predominant method of feeding is *ad libitum*—at one's pleasure.

Their skeletons deform under the load of their muscle mass. These deformations—osteopathologies including rickets (a softening and weakening of the bones usually found in children) as a result of vitamin D deficiency, degenerative joint diseases such as arthritis, swollen joints, bone-rubbing-on-bone, and twisted legs—are already becoming a distinct geological characteristic of the twentieth and twenty-first centuries, along with plastic and concrete. This reality hems me in, crookedly.

...

I try to write a lecture about the colour red—blood, wild strawberries, shame, 'meat' chickens, Mary Shelley's *Frankenstein*, Anne Carson's *Autobiography of Red*. I end up writing about blue. *My* blues, which at times feel so dark they turn violet to black, which are colours associated with regions of the deep sea, where sunlight cannot reach, and is often dismissed as devoid of life even when astounding creatures are found and brought up from the depths.

Blue is associated with edges, the unknown, with beginnings and endings. Maybe I am thinking about this colour because everything feels saturated with loneliness and sadness, which has created a climate or atmosphere of the blues.

...

Weather-making is counted among the chief abilities of witches. Historical climatologists are starting to pay attention to the idea that in early modern Europe it was thought witches worked in demonic conspiracy with the weather, that they indulged in weather magic and were burned for the sudden onset of frost, storms and hail, making crops fail.

In the sixteenth and seventeenth centuries, a rise in extreme weather—heavy rainfalls that poisoned fields, spread disease through cows, caused rising infant mortality and the outbreak of epidemics—was followed by a rise in executions.

...

*Christ is coming.* In 1806 these words appeared on the freshly laid eggs from a hen in Leeds. The hen was the property of Mary Bateman, known as the Yorkshire Witch. As a warning of apocalypse, these eggs caused panicked terror.

People came to see these egg-prophecies until it was found that Mary was writing the messages in corrosive ink and reinserting the eggs into the hen's cloaca. Eventually, she was hanged for murder on an unrelated charge. To this day, the Prophet Hen of Leeds has an active Facebook page. Thirteen likes.

...

Medieval scribes wrote on animal skins that had been salted, washed and cured into parchment. In Denmark, scientists working with book conservators have discovered a way to genetically analyse the skins. They want to do this to learn something of medieval life. A Gospel of Luke which dates from the twelfth century is a veritable menagerie—calves, sheep, goats and two species of deer.

...

Not just witches and weather and skins, but poetry makes history. The skill in crafting a poem translates to a skill in crafting the world. How will future geologists reckon with the fact that manifestos and poems have reconfigured the biosphere, alongside chicken bones?

## May 20, 2021

A cow comes walking along the road searching for some nice, sweet grass. Or looking for anything. I stand in the middle of the street telling myself *a cow is really walking down the road*. Then another and another step out from a paddock through a garden and into the street. There are seven. There are nine. Everyone runs out of their houses, waving their arms and shouting. The farmer roars around in his Ute, herds them away.

...

A wind comes up. Rain pitches down. Eventually, the wind will gather all the cows on a long truck and whirl them off to the last place they'll ever visit.

...

L and I feel terrible for those cows. We go inside. The sun burns to a smoulder. We bake pies—lentils, mushrooms,

potatoes. We stare out the kitchen window. I try not to
blink. Why keep watch? What else is there to see?

...

The night is an ocean. The immediate problem is keeping
warm. I drink red, white, rosé, beer, gin, whisky. Forty-
four drinks in the space of some days. I drink until my
breasts ache, till the lymph nodes in my left armpit
swell. I dream about illness.

...

We don't talk about the cows after that day, but night
after night when we slip into bed, a cow sloshing heavy
with sorrow lies down with us soon as we turn out
the lights.

...

I fear falling into the ocean. Of being thrown into the
ocean. The lights are out. I am falling. Then jerking
awake. I listen out for possums, for cats. The darkness
says nothing. There is only silence and sloshing. I
notice how the silence feels hostile. I try to translate the
hostility. But I am no translator. I try to enter sleep. I try
to enter memories. In both instances I am a traveller. I
am police checked. Must pass through tightened airport

security. I have all the right documents for sleep. Sleep is the terminal I am looking for.

I hold my documents out. I am told to leave my phone behind so I leave my phone behind. I am accompanied by an attendant who warns, *clear skies do not mean safety.*

I reach a point of departure into memories or sleep. This I begin to believe is a high-security moment. I flash my papers. I know I have no choice. I do not move but will be moved into one or the other: memory or sleep.

A woman is seated behind a screen. She gestures for me to step forward. I step forward, hold out my papers. I speak through the glass. Make my requests. Words feel heavy in my mouth. Heavy enough to crush my jaw.

The woman's eyes glaze. She gives a stale airport sigh. Looks at the enormous clock that is the sky. Looks back at my papers. Waves me through. Crossing over, I step forward on frayed feet.

Inside my memory room a translator is busy working in the corner. I look around, exhausted. All those stacked files. The translator's fingers are typing. Her head is nodding. I am bloated and quiet asking what, exactly, is happening now.

...

2013. A faded green slaughterhouse truck drives from Brooklyn to New York's Meatpacking District. In the back, over sixty cuddly puppet pigs, chickens, sheep, horses, geese and cows squeak, crow and cry. The side of the truck is painted with the words Farm Fresh Meats and a phone number to access an audio guide to the work. *This is a piece of sculpture art* a voice says down the line, *and I know what you're thinking, isn't it a bit subtle?*

The footage of the truck is taken on wobbling smartphones. Curbside kids jump and laugh and wave at the animals who cry out for help. This is Banksy's *Sirens of the Lambs*.

1985. In an interview, Georges Franju explains that by mistake *Le sang des bêtes*, his film about Parisian abattoirs, was shown during a children's matinee at the Venice Festival. Apparently, it had the kids *in stitches*. Later, Franju showed *Le sang* to a group called the Young Animal Club and they, too, reportedly, roared with laughter.

Banksy's *Sirens* is a moving reality fiction, a form of mediation that relates people in the street back to farmed animals, to abattoirs. It breaks the spatial construct of the abattoir as the repressed by literally steering it back into the buzzing ring of life. Like Franju's film, *Le sang*, *Sirens* is a hard truth made aesthetic.

# Bright Unbearable

Blue has made its way into poetry and literature so many times as the colour of exile, eyes, pencils, distant mountains, emotion (particularly melancholy), looming horizons, the future or end of worlds, of love and lust, of impossible places, and of oceanic landscapes, the salty edge of land. But as poet and essayist Eliot Weinberger writes,

> Go back far enough and there is no blue.
>
> Blue, black, blonde, blaze, the French blanc, and even yellow all derive from one proto-Indo-European word: *bhel—that which is shining, burning, flashing, or that which is already burnt.
>
> Homer's sea is notoriously wine-dark. Odysseus's hair is the colour of a hyacinth. (Milton, in turn, blind and a classicist, gave his Adam 'hyacinth locks').

. . .

The abattoir specialises in separations. Skin from flesh, flesh from guts, eyes from head, cheeks from jaw.

To try and understand this, I spend a morning watching YouTube videos of cows fighting crazily against it all. I learn that blood is not wine-dark but bright, even when spattered on floors, aprons, walls and across hairnets.

The word that comes to mind is *enargeia*, which Greek writers used when gods came to Earth as themselves, rather than in disguise. From Alice Oswald I learn that it means something like *bright unbearable reality*. Yes, I think. Yes. The blood in those shaky YouTubes is still alive and kicking.

What does it mean to watch a series of executions and then just go on with your day?

I leave the house and walk for a long, long time. I walk west to the setting sun. Then east to the moon. I walk for many days and many nights. When I return, I pour myself a glass of red. Then another.

I am not conjuring a myth of the poet who spins her excess sadness into story. I am wallowing. I am getting drunk as remedy. I am confessing to you the shapes of my booze wrecked flesh. I am joining those who make their sadness public.

I go sit on the couch. I am a stone of sadness. I swell and swell. I hit myself on the head. I drink, I hit. I am horrified to tell you that I am working to dull and amplify sensation, while sitting on a couch. I am trying to grieve better. Trying to navigate the infinity between wanting and doing with sharper instruments.

# On Red

1. A friend told me her father worked the abattoirs as a bolt stunner until he fantasised turning the gun to his own head. I imagine life and death pouring back and forth between his hands like hot lava.

2. I imagine some fantasies of death have geothermal pressure.

3. Geryon is a winged monster who lives on a fabulous red island and possesses a herd of cows. The cow's coats are stained by the red sunsets on Geryon's red world, with its red breezes that flow over red landscapes like red silk, like measures of wine, like a haemorrhage.

   Until he was killed by Herakles, Geryon had red cheeks, hot dry arms and was companioned by a little red dog. And I suppose he was still this red even in death. It is written that the end for Geryon and his pup was hateful.

4. In the story of Genesis, Adam and then all the other animals are created out of the same blood red soil. This means they are consanguineous. But then God grants

dominion, Adam names the animals and the power of naming cleaves.

5. As bird watcher Tim Dee notes, *'To cleave' is a verb that all taxonomists must fear, because it means to split and to lump, to pull apart and bring together.*

6. Because of Genesis, red becomes the colour of total creation and total destruction, of appearance and disappearance, of essential likeness and essential rupture. What lively dangers this story has gifted the world.

7. Red is the shadow biography of blue. And vice versa.

8. Blood is a beginning, even as it is so often taken only as an end. I want to write about blood to feel less alone in the hands of my vulnerability.

9. Geryon lives in a riot of colours. Roses roar. Silver starlight crashes against windows. Blades of green grass click. From what I can find, ghosts say nothing. They do not even appear, though that text is haunted. If they did appear, would they be heard as the crack of fresh white sheets or blue's muttered apologies?

10. The detective, the villain, the crime scene, the locked room, the spree, red herrings, investigations, false solutions; these are the features of stories of crime and investigation. Before French critics coined the term crime noir, cinema in the United States was serving films characterised by silhouettes, fog, guilt-ridden characters and the idea that *anyone* can become a criminal. As a bold new trend, it was originally called the red meat crime cycle.

11. I wake up one morning, and another. I stare at my red face in the mirror. My mouth a parted bloom. Inside, it is dark as a pond.

# Even Grief is an Immunitary Defence against Animality

When Koko the celebrity gorilla was first shown a skeleton, she was asked if it was dead or alive. She responded, *Dead, draped*. Draped meaning covered up, buried. When asked where animals go when they died she replied, *a comfortable hole*, and mimed kissing the world goodbye.

It has been thought animals don't have the capacity to visualise their own deaths. This has been used to justify the idea that they cannot take their own lives. But elephants have walked off cliffs, trampled their own trunks; whales have beached, intentionally.

Dogs have jumped out windows and off bridges into the enormous unknown, and never returned. In 1845 an article reported that a Newfoundland dog leapt into a river and refused to swim. Each time he did this he was rescued. Each time he was rescued he jumped back in. Eventually, as the story goes, he just stuck his head underwater till he drowned.

...

Lately, when I get into a car, my mind plays out bloody scenes. A truck. A tree. A verge. All become agents of death. When I stand at the train station, cross a road, chop vegetables or take up the axe for firewood, a wild profusion of thoughts about death shake like winds, rattle like snakes, roll around in my mind like angry stones. These visions gather slowly or quickly. They happen in multiple viewpoints. Always I am the victim. But sometimes I am also my own vampire witness.

...

Henry Maudsley, author of 'The Genesis of Mind', argued that animal-lovers misinterpret accidental animal deaths as deliberate suicides. In the face of growing claims for animal intelligence in the nineteenth century, Maudsley pushed the idea that even in fits of grief, despair or mania, humans are superior to all other species.

According to Aristotle, a horse once leapt to his death after learning he had been tricked into mating with his mother. Romans saw animal suicide as natural and reported many cases of horse suicide. Tales of scorpion suicide have long featured in Iberian folklore, and in 1813 Byron repopularised the image in his poem *The Giaour*:

>    *In circle narrowing as it glows,*
>
>    *The flames around their captive close,*
>
>    *Till inly search'd by thousand throes,*
>
>    *And maddening in her ire,*

*One sad and sole relief she knows,*

*The sting she nourished for her foes ...*

...

Early this morning I met death. He was standing on the low branch of a eucalypt. *Are you death?* I asked. *Yes,* he said. *But don't worry I don't have my scythe or anything. OK,* I said and walked on. *Are you kidding?* he shouted. *You're really not going to take a photo?* He posed. *I don't have my phone,* I said and opened my hands in a gesture of helplessness. *Awwwww, that would have been the first photo anyone had taken of me in years.* He hung his head then jumped in front of a car. Later, a boy with brown hair ran through the street shouting *Hello! Hello! I'm Death! Remember me?*

...

Norwegian lemmings have been shown marching in masses down from their mountain homes to the coast, where they supposedly drown in the sea. Disney's film *White Wilderness* initiated the legend of these suicide events. Turns out, though, the lemmings in this film were forcibly marched to their deaths while the film's narrator described their unreasoning hysteria, that they move on ... move on ... move on ... away from the shore, with no more thoughts of food or life but, actually, the Disney people bought the lemmings from children in Manitoba and made them run on snow-covered Lazy Susan-style turntables before throwing them off a cliff and into a river. The dirt of this fact gets right in under my fingernails.

In *A Year of Magical Thinking*, Joan Didion writes, *Grief, when it comes, is nothing like we expect it to be.* So why is it so often described as a relationship, a traveller, an uninvited guest, a tunnel or shaft of bright, unbearable light? Is it because we can't outrun this kind of misery that we try to squish it with banalities?

. . .

In the nineteenth century, the medical and psychiatric debate of whether animals were capable of suicide, or mere self-destruction, exploded. It was decided that experiments capable of replication were required to prove whether grief, dread or mistreatment could drive an animal to take their own life. After Byron's poem, scorpions featured heavily in the experiments.

. . .

Tonight I'm surprised to find a full moon sitting on the far rim of the sky. I ask the moon, *Why are old men, young women and even houses thought to carry more souls than other animals?* As if they were empty as empty fruit bowls? Soft blue moonlight glides across the outer world and cold neon tracks cross the kitchen floor. This indecipherable response slings my mind into a misery.

In the sad story of Gombe Stream chimpanzees, as told by Jane Goodall, the chimpanzee called Flint developed signs of clinical depression after his mother, Flo, died. At first Flint stopped spending time with his family. Choosing, instead, to lie on his back near to where he last saw her. His eyes sank into his head. He stared into the distance, sometimes without blinking. He refused to eat. No turning back. Lost a third of his weight by the third week of fasting and slid out of himself the very next.

What grief is this? Kidnapper? Bully? Prison warden? One of light's beaten moods? A pair of wings, unclipped? Whatever it is, it is. It comes one afternoon. And then another, and another. A shiver of something. It falls, straight from the eye of life.

To record the dying body. The dying body, which always happens in the middle of life, *in media res*. Must we write life dyingly? as Christopher Hitchens said in *Mortality*. Or, write dying livingly? Or, write a world that is just like this world, only different in a very meaningful way? A world filled with sad claims of impossible things.

Primates are known to carry, groom, inspect, protect and peer into the eyes of their dead children. Rousseau has said *an* animal will never know what it is to die, and the knowledge of death and its terrors is one of the first acquisitions that man has made in moving away from the animal condition. Philosopher Vinciane Despret offers a way to think about Rousseau's conceptual moves here:

> History shows us that what is at stake in these conflicts of attribution of sophisticated competencies to animals can often be read, if one may forgive the barbarism, in terms of 'propriety rights of properties': that which belongs to us— our 'ontological attributes,' like laughter, self-consciousness, knowing that we are mortal, the prohibition of incest—must remain our own.

I go looking for the kind of poetic language that can thaw ice or make the dawn appear and be used to describe the after-death rituals primates show their loved ones. Elephants, who pay homage to the bones of their dead, have been allowed to share in the word *reverence*. But common chimpanzees, who may stop carrying the body of their dead beloveds after a few days, are only given *abandon*.

...

The story of a Southern Resident orca, Tahlequah, went viral after she was found pushing her deceased baby for seventeen days straight through the Salish Sea off the coast of San Juan Island, Washington, the lands and waters of the Lummi People.

She did not eat for seventeen days, she just swam. Sometimes she carried her child in her mouth. Sometimes she pushed her daughter with her rostrum. She was accompanied the entire time by her six-year-old son. Her pod remained close by. The world wept. And then, I think, the world forgot.

. . .

Do you know that illustrated book *Duck, Death and the Tulip* by Wolf Erlbruch? One day Duck turns around and finds Death standing behind her. Duck is frightened. *What are you doing creeping behind me?* she asks. *You've come to fetch me?* she asks. *Oh, I've been close by all your life—just in case*, Death says. *In case of what*, Duck asks. *A cold, an accident*, says Death. Duck is suspicious; she has only just noticed him. But Death is friendly, it is summer. They go to the pond. They climb trees. They are, simply, living. Summer turns to autumn and one evening Duck feels a chill. She asks Death to warm her, so he wraps his arms around her and soon Duck stops breathing. In the morning, Death finds her lying quite still. He picks her up and carries her away so lovingly, so gently. *That's life, thought Death.*

. . .

Some experimental psychologists designed tests *sufficiently barbarous to induce any scorpion with the slightest suicidal tendency to find relief in self-destruction.* These tests included burning them with acid, alcohol, condensing sunbeams on their backs, heating them in bottles, exposing them to electrical shocks and other *sources of worry.*

Did you know scorpions are arachnids? When I think of them, I imagine them coloured in bright cartoon reds. And though

I am hooked on scorpions, their sources of worry, the colour blue is eating my mind: the blue of manta rays, the mineral shimmer of ghosts, of water, deep ice, full moon light, the Salish Sea, arteries, the colour of cartoon raindrops, the colour of certain moulds, country music and Samantha Hunt's novel *The Seas*, which is filled with oceanic blues, drowning love, drowned language, depression, alcoholism, mermaids and PTSD.

Did you know it's criminal to be a mermaid these days? If you kiss a human, they might drown. They can drown even if you kiss them in the middle of their kitchen.

Mermaid (n.) a woman with no fear of depth, as Anaïs Nin once said.

Or a woman with strange attachments.

A sixteenth-century illustration of a mermaid-merman pair by Italian naturalist Ulisse Aldrovandi shows the merman reaching his arm round the shoulders of the mermaid, his hand fishing for her right breast. And did you know that Princess Ariel was groped by an accused sex offender in the exact same way while posing for a photo in the Magic Kingdom? Ariel's Grotto was temporarily closed.

Mermaids are not thought to have a soul until they are married. A mermaid was donated to the Smithsonian Institute in 1890, is said to have had a well-developed bosom and is supposed to be kept floating in a six-foot glass jar of alcohol.

It is no accident that drinking is linked to drowning in the sense of getting drenched, swallowed up, engulfed, submerged, soaked, saturated.

The mermaid in Hunt's novel *The Seas* lives on dryland in a town with the highest rate of alcoholism in the country. A wetland in a dryland. She is in love with a soldier who has returned from Iraq.

The mermaid loves the soldier like dead loves alive, like wet loves dry, like sea loves sky. She loves him so much she almost goes blind. The soldier loves drinking so much whole bottles of Canadian whisky come get him like king tides.

...

Blue is filled with wine darkness. Sometimes I think that blue is full of everything except red.

...

Since the 1950s over six hundred dogs have leapt from a single spot, on a single bridge, in Scotland. The bridge is nicknamed Dog Suicide Bridge. A local teacher believes the bridge holds a menacing spiritual quality, as the site of a murder and attempted (human) suicide. A place where life and death overlap. Once, the teacher stood at the spot in question and experienced a strong jab like a ghost finger.

...

Lab experiments on animal suicides do not appear to have taken the presence or absence of ghosts, mystical visions, residual raptures or the collapse of distance between earth and it's celestial beyond into consideration during their cross-examinations of multispecies depression, mania and self-destruction.

. . .

Tonight, my dreams come in the shape of a cemetery. Hundreds and hundreds of plots.

# Hauntings Come . . .

... when the over-and-done-with rustle like grass in a breeze ...

... for the French philosopher Derrida, the traditional scholar does not believe in ghosts but there is another kind of scholar who *is* open to spectrality, they're not someone who trusts in the return of the dead, but thinks the spectre as a possibility, who thinks of the mixing and overlapping of non-life and life, this other scholar is a Paranormal Investigator ...

... 'para' being a prefix with the meanings of 'at or to one side of', 'beside, beyond, past, by' and 'at the fringes of the normative' ...

... dominant research practices and processes marginalise the Paranormal Investigator, as such, their work might fail to be understood within traditional research conventions and uses and did you know that to go paranormal in your research is to glean from the mess of life all those things that have been trashed by the orderly structures of common sense or hegemonic knowledge ...

... when you come into contact with a ghost, you are being notified of what is concealed but remains vitally present and the Paranormal Investigator takes note ...

... Kafka was a Paranormal Investigator, like, in *Ninety-nine Stories of God* Joy Williams writes: *At some point, Kafka became a vegetarian. Afterward, visiting an aquarium in Berlin, he spoke to the fish through the glass. 'Now at last I can look at you in peace, I don't eat you anymore'* ...

... this makes me think Kafka had a *fleischgeist*, which is a German term, a made-up word that means 'flesh ghost' and 'meat spirit,' to have a *fleischgeist* is to have the horror of one who has been turned into meat transported, and felt, in your own bodymind ...

... imagine pale mortuary animal figures roaming city streets and highways, entering drive-thrus, bellowing lamentations down intercoms ...

... the poet Jack Spicer insisted he did not write but dictated his poems, said he was a receptive host for language, that he received messages from ghosts, he would stay up at night turning their transmissions into poetry ...

... he died at forty, from alcohol, and there's the Fox sisters from Hydesville, New York, are said to have birthed the spiritualist movement when they claimed to have made contact with a murdered peddler, this was in 1848, they suffered a strange time of bangs and raps in the walls at home, the youngest started to ask questions of the noisy visitor ...

... responses came in bangs and raps and from this conversation with the spiritual realm the sisters became famous then they were disgraced and by the mid-1850s both were alcoholics, they died penniless, apparently, it's common for mediums to have addiction problems ...

... ghost stories are not stories to simply pass on, not campfire fodder, they are the cries of those who have been made killable returning to confront their injustice, who will confront it with them?

# Notes on Ghosts

Ceridwen Dovey's *Only the Animals* recounts the last days, months, years of multispecies lives. There is the voice of a dolphin, Sprout, trained by the American Navy Mammal Program to find underwater mines and traps, and to mark them by *dropping acoustic transponders close by*. There is the soul of a camel, killed on a wild colonial desert trek. There are souls of an ape, a tortoise, a cat ...

There is a dog, too. The soul of a German dog who finds himself on the battlefields of World War II. He is lost. Wandering. Hungry. Picked up and *drafted into a legion of dogs who were to be given the special honour of leaving the camp to accompany the soldiers into combat.*

Day after day other dogs—dogs that are starved, *going mad with hunger*—are loaded up with a special heavy metal pouch with a *metallic pulse of its own* and sent outside. One day it is his turn. To make him run he is pelted with stones: *I set my nose to the ground to try to find my way back to the Germans, to my compatriots, hoping one of them would risk everything to save me, and get the metal blight off my back.*

He never does make it.

A note on ghosts: they do not wait to be invited in. They reach up from beneath floorboards and grab you.

Try to repress a ghost and they will erupt back onto the social scene, for the paranoid mantra that says *ghosts are not real* cannot be maintained with the required force to keep them offstage forever.

Is seeing ghosts a failure of social etiquette in which we must all commit to collectively and continually deny traumatic or horrifying events?

Do streets and libraries, supermarkets and houses really need to be spiritually cleansed? Is an energetic detox a heroic act?

Everywhere I go seems to be the exact wrong place to ask these questions. But I carry them with urgency, I race around holding then like they were organs in need of a new body.

Critical race studies scholars Eve Tuck and C. Ree write that official logics seek spectral containment while paranormal logics understand the mutual implication of the living and the dead, the remembered and the denied. They write specifically about the ghosts produced by settler colonialism, which is the *management of those who have been made killable, once and future ghosts* ... They say that the process of 'making killable' *turns people and animals into always, already objects ready for violence, genocide and slavery.* To make killable is to make subhuman, to animalise, to meatify.

For a week I try practising a kind of ghostly envisaging. I stand at the back fence and gaze into the paddocks. I stare out the kitchen window, mid stir-fry. But I don't see anything. I don't see anything. Just the darkness. Just the light.

Then, as if a spring broke loose, a probing light appeared. It was blue, carried a weight of intellect and self-regard. First it was in the doorway to my house, then off between the trees and roaming the edges of the paddocks over the back fence.

In a haze of sadness, I sit on the couch and watch the minutes burrow through the meat of the hours. I should be writing a lecture. But I can't ring a sentence out. Minutes gorge on hours, till all the dead and infected tissue of my day has been cleaned up. I sit very still. Trying to summon up anything from anything.

I learn some basic paranormal cleaning tips online. You can burn white sage, white candles, fill rooms with white roses or scatter black salt. I don't want to detox my house of its spirits, only I'm surprised to learn how particular ghosts are about colour.

I walk through the house burning a stick of incense. I want to summon, not banish, ghosts. Though it seems I've only been able to call forth a family of mice who come in the dead of the night and nibble our bread.

To be seen by a ghost is to be looked upon from another horizon. To align yourself to the view of another. To be understood or misunderstood. To be held instead of beholding.

A glass tank is a frame, says John Berger. Say you notice a fish glancing sideways, attentive, wary, giving you a suspicious sweep. Say you notice how the unfree body has free eyes and you feel, perhaps for the first time, the tank is framing you. What Berger says of zoos is also true of aquariums, they are *epitaphs to a relationship* ...

Looking is a tool. It is also, of course, a weapon.

Finally, I deliver my lecture on red, which is actually about blue.

Students stare at me through their digital windows. We are still huddled away from each other in sunless, interior regions.

My face is a bloated knot of pasta. Afterwards, I go sit on the couch and chase my grief. My solid blue-black grief. Dirty apartments of grief. I pick through the permanent squander of it. What am I doing? Grieving the deficient meat of my grief.

November 2021

A student writes about psychedelic drugs. My mind drifts to thoughts of trepanation, that ancient technique of drilling or scraping a hole in the skull, of making intersecting cuts, scraping, grooving, boring to relieve pressure, expand consciousness and enhance the effects of psychedelics. In 1970, British countess Amanda Fielding made a film of her first experience of auto-trepanation: the self-cutting of a skull.

These words—the self-cutting of a skull—have a place in any conversation about writing. Don't we turn up to the scene of writing, each of us wobbly bags of water, held together by thin skin, an eggshell skull, and make incisions? Isn't that what we are doing together and/or alone when we wake up and write, come to class and write, pull over in the emergency lane and write? Aren't we cutting holes in our skulls to extract memories or dreams, questions, confusions, anxiety, fears, joys, madness, light and shade, traffic and offences, travel at night and unfamiliar places, like stones? And don't we lay those stones beside other stones? Stones that have come from places far away or close by. Moon stones. River stones. Sand stones. Stones we find, *there!*

beneath our feet at the beach. And then? You start with a mess of stones. You arrange the stones. Rearrange the stones. Take some away. Break others apart. Pile them up. I want to say if writing is earth moving then when writing shakes, everything else does, too? That you, the writer, will spend your days leaping from cliff to cliff? That, mid-flight, everything—all your stones—will fall out of your pockets?

Months ago I searched nervously for Fielding's auto-trepanation film but could find only snippets here and there online:

1) Fielding tapes sunglasses to her head to stop the blood running into her eyes.

2) Fielding cuts her fringe.

3) Fielding tapes back her hair.

4) Fielding puts on a shower cap.

5) Fielding holds a small electric drill up to her forehead.

The snippets feature a pigeon named Birdie. Fielding rescued Birdie as a chick and I learn that they share a telepathic bond. That they are lovers.

...

There are days when I feel that I have to write all this down quickly, before I lose too much blood. Writing as if writing will cure something that exists, undiagnosed, in my body. Writing through nausea, back pain, head-aches, breast ache. I would like to see a doctor except for the financial pressure of it. I write through nightmares that I have developed breast cancer. The essayist Kate Zambreno has had these nightmares about breast cancer, too. Separated by age, time, continents and so much else, we both wake at night terrified of medical bills. Real/imagined. When I wake, I turn on my phone's torchlight and slip into the memoir of her leaking, bursting, time-bomb-of-a-body.

...

I get the feeling that tiny pieces of glass are stuck in my throat. Writing does things to every body. You are isolated on an island. Your skin is peeling off. You write like you *are* an escalating fever. It, writing, can feel invasive, diagnostic, you swell and then it drains. Writing does not shape, but tracks truth beneath your skin. Writing drills holes into language so language can expand; writing should let old blood empty out. As language expands your ankles grow weak. Sometimes your skin erupts.

...

What can writing do for other bodies? This is one of the questions concealed in the shadows of Zambreno's *To Write as if Already Dead*, which is a twinned study of Hervé Guibert's final novel, written as he grew increasingly ill from AIDS-related disease, and

Zambreno's pregnancy, her various illnesses, writing, exhaustion and financial fatigue. Zambreno writes in moments, mostly in fifteen-minute gaps (an hour if she's lucky) between the demands of the living.

...

Sometimes I think I am already dead, that these are the thoughts, questions and feelings of a ghost. Atrocious things have happened to my body. Mostly, I envision a car crash. Being cut from steel. I carry these ghost images with me up and down the highway.

...

I am one year older than Hervé Guibert was when he died, which was two weeks after he turned thirty-six. I can't tell if these are petty confessions or intimate revelations of what plagues my living meat.

...

In poetry a caesura, Latin for 'cutting', is a metrical pause or a break in verse where one phrase ends and another begins. It can be expressed by a comma or two lines //. In music a caesura is a pause, a silence, time not counted and it is represented on sheet music by two slashes, train tracks.

Jenn Ashworth's memoir, *Notes Made While Falling*, uses the caesura // to designate each time a new, untellable, story begins. A beginning, she writes, *is a cut in the onward flow of things.* The caesura reminds me that even while Ashworth has called so many words to the page, the narrative really, somehow, remains hidden. A shadow.

The caesura's slash signifies the place where words, sentences and phrases have dropped off the page, go missing in action. It holds the place of erasure, absence. Shows writing as a leaky container, filled with holes. The caesura // signifies derailment.

Ashworth's deep need to explore cuts and cutting originates with complications that came after her c-section delivery, when she was rushed back to the surgeon's table and opened up, again. During this operation, her epidural started to wear off. Paralysed but awake, panicked, she felt things beyond words. But she put words to them: an extreme pressure, a wind tunnel blowing through her organs. Her legs were rubber. No. They were dream legs. She needed to move these dream legs to convince the surgeons that she could feel what was going on. Speaking of her trauma she writes, *That's what I've got to do. Make a book out of it. Make a book out of it. Cut a hole in a life and through it, pull out the story. But I can't.*

Ashworth's entire memoir consists of drawing significant philosophical, aesthetic and ethical connections between different kinds of cuts. Each one bleeds into another. Like Alice's rabbit hole is also a wound in the chest, is a mine, is a cave, is a hole in the head made by trepanning.

...

Pain cuts narrative to pieces. Narrative—porous, uncertain, temporary and sometimes terribly sick—can also put pieces of your body, and others, back into some kind of together, again.

When I first learned about caesuras, their use in poetry, I repeatedly made the mistake of writing that the slash signalled a juncture between worlds. *The caesura is the point at which one world ends and the next begins*, I wrote in an essay for my second-year poetry class. It should have been *word* not *world*. But I hold onto the mistake because it feels truer, bigger, heavier, more correct than its corrective.

...

Have you seen the popular eighteenth-century woodcut from Normandy in which a Skull Doctor is, with hammer and anvil, forging brand new heads for women who have been brought for treatment by their husbands? This print is one of several variations satirising the feminist demands of the bluestockings of Paris salons, writers and poets. *I will make you good!* The Doctor says, hammering women's heads all the good day long. *Husbands, rejoice!* cries an onlooker.

And do you remember that scene in *Hannibal* when Hannibal Lecter cuts a section of brain out of an FBI agent who is fully awake though drugged and sitting up at a dining table? Well, I've learned that there is a way that this is both possible *and* untrue.

Though the brain contains no pain receptors—nociceptors—it is the organ through which you interpret and evaluate thoughts, feelings, relationships. It is also said that physical pain, experienced with enough intensity, obliterates psychological pain.

I need to grab onto these facts because I need examples of the ways one can be both totally fine and dying.

# Anatomies of Force

When Virginia Woolf meditated on the roots of war in *Three Guineas*, she looked at a photograph of a freshly bombed street and saw that it contained a confusion of flesh, a *zone of indiscernibility*, that *might be a man's body, or a woman's; it is so mutilated that it might, on the other hand, be the body of a pig.*

In *Regarding the Pain of Others*, Susan Sontag reconsiders Woolf's description of bodies meatified by war and notes that when ruin is so thorough, one can't always make out the subject. Indistinction, itself, becomes a weapon.

A dead body, writes Simone Weil, is a thing. The power to kill comes only after another, more momentous, power—that of making the still living into a thing. Even before being touched, one becomes a corpse, or a hybrid living-corpse. Weil writes this specifically about Homer's poem *The Iliad*, which for her is a poem of force, of the living as corpses, whose primary concern is the force of brutality. Force is the poem's centre.

Weil's anatomy of force includes:

Force that dominates. Force like excessive hunger. Cold, hard force that governs all matter. Force that possesses those who believe they possess it. Segregating force. Force of tears. Force of fate. Force that renders misery incomprehensible. Force of revulsion. Force that effaces inner life. Force of silence. Of refusal. Of fear.

Force that eliminates the grief of a mother. Intoxicating force. Humiliating force. Force of trembling. Force of a river. Of blind destiny. Of the gods.

Force possessed by those who believe in mastery. Force of justice. Force of no reflection, hence ruthless and mindless violence. Force of flowers cut for a grave. Force of injustice. Force of silencing an old man with a single word. Balance of unequal forces. Abuse of force. Of want. Of the mundane.

Force of a house in ashes. Force of a single detail. Force of a few brief hours. Of taking the sea, the whole sea. Force of an empty head. Of a secured departure. Of empty hands. Of life.

Force of begging for your life. Of being on your knees.

Force turns a body into a thing through war, says Weil.

The point is to out-injure another; to alter the body tissue of the world. To force another out of place, or out of life, itself.

A stick, a stone, a hand, a door, water, a knife all become weapons in the world of war.

Not all force is created equal. Just as Maggie Nelson writes *not all 'thingness' is created equal.*

Each body has its own historical relation to objectification, which can be called *thingification*, which can also be called *meatification*. Each hand has its own historical relation to force.

When flesh is beaten with bald cruelty it is called meat. For years I have wanted to write unapologetically that slaughter is the product of war, that war whets the ground for slaughter, which whets the ground for more war, that I am not speaking in metaphors.

# Cannibal Cafe

*We are food*, wrote Val Plumwood, the philosopher once seized by a Saltwater Crocodile, the largest of the living saurians. She learned that *to be or not to be* meat is a thin red line, not a gulf—a matter of perspective, not a matter of fact.

...

Of course, says Francis Bacon, the painter of all the anguish, suffering and liveliness of meat, *we are meat, we are potential carcasses*. According to Bacon, even those earthly creatures who have been denounced and defaced will not stop calling 'us' to the reality of our *own being toward meat*. Say you give into this recognition and enter the terrain of meaty democracy, a system governed by blood, mucus, jaws, knives, boreholes, nerves and pulse.

...

Now, it is a legendary catastrophe. Then, it was just a day in February 1985 when Plumwood went canoeing in a place where the East Alligator River surges in the Stony Country of the Arnhem Land Plateau. She encountered a branch in the river where food and death met. Up rose a crocodile. She looked into the crocodile's eyes before she was taken down. For decades she would sit with that look, try to reckon with the saurian point of view. The one that saw her body as meat.

She was taken for three death rolls and survived. Tenderised but alive, she clawed her way out of the river and later wrote that to deny the foodiness of humans refuses the perspective of those who position us as cuisine.

Meat is a standpoint, a perspective, and thinking from this perspective is like travelling to another place, towards another horizon. It means getting lost to the feeling that you exist outside the edible earth community. I sit back and try to comprehend this, but my mind is an unoccupied seat.

...

So much narrative is driven by a desire and repulsion of appetite. Take the story behind the story of Hansel and Gretel. It is about two rival bakers in Nuremberg, Germany, 1618.

> One baker makes the best gingerbread in the mountains but won't share her recipes.

> She won't be tricked into marrying her unworthy rival, who wants to have and to hold her recipes for better or for worse. But mostly for better.

Her rival spreads rumours about her. Calls her a witch, which it is a capital offence to be.

Her life becomes unliveable in her mountain village. She flees. Goes to live a solitary life in the woods. She abandons everything but her recipes.

The rival baker and his sister follow. They break into her house. They murder her. They burn her remains and ransack upstairs, downstairs, kitchen but can't find the recipes.

The murderous souls eat all the gingerbread left in her house.

Which is to say so much narrative about appetite is driven by desire, greed, harassment, jealousy, moral bankruptcy, coercive control, false accusations and rumour and ends in murder.


...


There is another story about appetite so terrifying it does not need to resort to the supernatural to haunt listeners. A woman and her two daughters are starving. The woman tells her daughters that she will have to kill them to satisfy her hunger. They appease her by begging for bread.


...

Melanie Klein, a disciple of Freud, pursued the possibility of hostility in the mother–child relationship when she wrote that the mother's breast, first a source of gratification, becomes a target for destruction. Years later, Susan Suleiman would point out that Klein never spoke of the murderous impulses a mother may feel towards a child. I do not have a child but if I did, I fear how much I'd fear a mother's position in this homicidal double helix.

...

Abandonment, cannibalistic hatred and disfigured forms of parental and sibling aggression come into view in so many folk and fairy tales. A stepmother cooks her stepson and feeds him to his own father. Cannibal brothers consume their sister. A cannibal sister tries to eat her brother (but he runs up a tree and so lives). Chronic food shortages feature heavily in these fairytales. *We can't go on like this*, someone declares, and a story begins.

...

The Japanese cannibal Issei Sagawa's favourite fairytale is, reportedly, *Hansel and Gretel*. Sagawa dreamed of eating someone for thirty-two years. And then he did. He describes himself in fairytale terms, as a weak, ugly and small man who ate a woman to absorb her energy.

Sagawa had been living in Paris, writing a PhD in comparative literature at the Sorbonne. He had been to a poetry reading with a friend, Renée Hartevelt. Afterwards, while she was sitting in his lounge room reading out loud, he took a carbine rifle from a drawer, pointed it at her back and pulled the trigger. The Rolling Stones wrote a song about it:

> *You know, he took her to his apartment / Cut off her head, put the rest of her body / In the refrigerator, ate her piece by piece ... took her bones / To the Bois de Boulogne / By chance, a taxi driver noticed him / Burying the bones, you don't believe me? / Truth is stranger than fiction*

Actually, Sagawa didn't bury the bones. He stuffed Hartevelt's remaining body parts into two suitcases and dumped them in the Bois de Boulogne, which was once the hunting ground of kings and is now considered to be the lungs of Paris.

Due to a loophole in Japanese law, Sagawa will never go to prison. Instead, he drinks iced coffees and strawberry milkshakes with journalists. He does restaurant reviews as a sushi critic and has starred in porn films. Being a modern cannibal, he says, doesn't pay as much as it used to. Now, he wants to die. Wouldn't mind being eaten by a beautiful young woman.

*So you would like me to eat you?* a beautiful young reporter asks Sagawa. *Yes*, he says. I wonder what kind of energy he thinks she would absorb.

...

Kathleen Knight, once a meat worker in rural New South Wales, is now known as Australia's Hannibal. She never actually ate the flesh of her partner but, like in a fairytale, she killed him, cooked him and served him on plates with gravy and vegetables ready for his two children.

Knight has been incarcerated for life and is known by fellow inmates as The Nanna. She is not a beautiful young woman. But she is, according to the biographer of her crimes, very well respected.

Biographer isn't the right word here. Thanatographer—one who would more specifically write of death—makes a better landing.

...

In 2010 Vebu, the German Vegetarian Society, publicised the opening of a hoax restaurant, Flime. It invited patrons to offer up body parts to be turned into gourmet cuisine. The restaurant's name, an acronym for *Fleisch Isst Menschen*, Meat Eats People, aimed to point to the personal and planetary injury that comes with the overconsumption of meat. But the hoax only served to jab at the festering wound sliced into Germany's psyche after Armin Meiwes, a computer repair technician living in the picturesque rural town of Rotenberg, now known as The Master Butcher, was found to have killed and eaten a young man he met via the online noticeboard for fetishised cannibalism, The Cannibal Café.

Growing up, Meiwes lived beside a self-professed witch who was once taken to court for casting death spells. In a police interview, he explains he had a childhood obsession with *Hansel and Gretel*. Brandes, his willing victim, is said to have harboured a corresponding obsession with this particular fairytale.

German Neue Deutsche Härte band Rammstein wrote a song called 'Mein Teil', which means *my part*, is slang for *my penis*, and is about Meiwes. Marilyn Manson used Meiwes as inspiration for his album *Eat Me, Drink Me*. Ozzy Osbourne released his song 'Eat Me' on his twelfth album, also inspired by Meiwes.

Today I've spent a good deal of time on the Twitter page and website of *Meiwes/Brandes the Musical*, which draws extensively on court transcripts and personal correspondence between the writers and Meiwes. The production's opening track, 'Raw', swims into my headphones, a sugary duet. It could be a love song, which the writers point out is the aim.

> *I've been waiting ... I've been wishing ... I've been holding out for you ... when we meet face to face ... will I still be your taste?*

...

When journalist Sandra Lee wrote *Beyond Bad: The Life and Crimes of Katherine Knight, Australia's Hannibal*, she made sure to write vividly of Knight's victim as a good, hard-working, ordinary man who loved his family, his mates, his work and his beer. Though I have been searching, I can find no songs or poems or vivid portraits about Renée Hartevelt, Sagawa's victim.

...

Human bones and flesh carry a special place in fairyology. Demonic spirits have a hunger beyond need and they live in a parallel world. Sometimes that world is made of candy. Tom

Thumb is fattened by an ogre. Hansel and Gretel are preyed on by a witch. The giant in *Jack and the Beanstalk* has a voracious appetite for human flesh. Baba Yaga is known for eating human beings as if they were chickens:

> *the fence ... was made of human bones ... the spikes were human skulls with staring eyes ... doors had human legs ... human hands for bolts ... a mouth with sharp teeth stood in place of a lock.*

...

In 2009 Travis the Chimp—raised as a semi-human child-pet by a couple in Connecticut—tore the nose, ears and hands off a visiting friend. He was fourteen at the time. He was also morbidly obese and agitated after a lunch of fish and chips and ice-cream cake.

Up until this day, Travis had slept in his 'mother's' bed, ate lobster at the table, drank from a long-stemmed wine glass, loved watching sports with his 'dad' on TV. To pay for his upbringing, his 'parents' rented him out for TV shows and commercials. On the day of the killing, his 'mum' slipped a Xanax into his tea because he was in such a bad mood. She called her friend to come over and help calm him down.

When Travis attacked his 'mother's' friend, his 'mother' began beating him over the head with a shovel. Then she stabbed him, repeatedly, with a kitchen knife. Police arrived and shot Travis in his own front yard. Wounded, he staggered, ran. While the police searched for him in the nearby woods, Travis slipped back into the house, his house. Into his bedroom. He lurched onto his bed and died.

# A Smack of Sorrows

Nine a.m. We're stuffed into a classroom together. The room has no windows so there are no trees, no birds, no cars, no roads, no river, no dogs, no people. Just us. Strangers in a little Enclosure. Introductions. *What sort of animal would you be if you could be an animal?* Rat, Cat, Marmoset, Octopus, Labrador, Human, then *death opened, like a black tree, blackly.* Plath from her poem 'Little Fugue'. So, Plath is here, too. And there are others. Writers, poets. Living and dead. And we will learn something from each of them. Or we will hate them. Or we will refuse to read them. Doesn't matter because right now the air is electrified. The poem flies free. Admire its dips and rises.

A crowd breaks out of the library in a confusing disturbance of shouts. The sound hits me like the noise of a riot. Laughter like a fist full of coins hitting the ground. A professor-type appears in front of student-types and she strides towards the library. I follow. Inside I meet Z the librarian who tells me about a new phone app in which cows peel back their bovine faces like hoods and converse with farmers. They make dire predictions about the end of the world. *I like cows* she said *so I like this game. I think I have a problem.* I try to find this app but the only one I can find makes me put hats on cows. No prophecies. Just lots and lots of hats.

We gather. We retreat. The world fills up with questions like water overruns a bowl. Campus is empty, again. Little signs have popped up like monochrome weeds. We are instructed not to breathe on each other. Out the window at my desk, absolutely nothing moves. It looks like everyone has gone, like nothing has survived. Not even a single bird. If this were true, how would I redeem the world? What words would be my amulets? I think about how I would bargain with Death.

Someone says, *Why are you still here?* pointing to one of our little office weeds. I rush to the train station, saying goodbye to each street corner, cafe and bookshop I pass.

At home I drink too much coffee. Lie on my bed. My eyes feel sunken. My flesh thickens out, and there are days when I am crying, or screaming, or totally at peace. In our glowing digital classroom, we are uncanny citizens. Old souls trapped in new wine bottles.

...

A friend emails me this short article about a suburban butcher who asked a fine-art nude model to pose in his meat cabinet for a photo shoot. The model agreed, thought it was a great idea. The End. I read the article over and over, thinking it might be a piece of cryptography. That there's an algorithm or secret analogy inside it. The butcher tells the reporter that the idea for the shoot was his *brainchild*. I devour hours whole trying to find some inner depth to this.

*A Thursday,* says L looking up at the sky. *A Thursday?* I ask. *It is a Thursday,* L agrees and goes to collect firewood. At sunset I watch for L's return. L comes in with a loaf of bread, some onions, lentils, olives and wine.

Back on campus, in the library. A tall student and a short student huddle together in the 800s—literature and rhetoric—and discuss their preparations for a deep pantry, which is something designed to outlast an apocalypse.

Theirs will be composed of beans, grains and nuts. To survive The End they will need to be at their peak physical game so tinned pudding isn't the best idea. They determine the goals of their pantry. Consider the location. Agree on whether they will build or buy shelves. Then they pack their bags and walk into their future.

I spend the afternoon responding to questions. What the voice is like. What the narrator is like. Speech is not the same as dialogue. Think of the last time you were involved in a minor conflict. Borrow a voice from real life, dissect its mannerisms. Try to think beyond the idea of an *arc*.

Outside a student asks me to take a quick questionnaire: *Describe your feelings of western philosophy in twenty-five words or less*. All the winds of the world swing around and batter the sides of my head. I can only think to write the German word *unüberholbar*, unsurpassable, which I think is another word for death.

...

It is now known that many aspects of planetary destruction are slow, unspectacular, cumulative and largely invisible, until there is a jolt, a slip, a tsunami. This presents a wicked problem for art. How to make the unspectacular watchable, readable, hearable?

...

At a reading group, a woman announces that plants are just as alive as chickens. She says she does not like to point this fact out in the presence of vegans. The room tightens several notches. I hold my breath.

Afterwards, a friend and I go for beer. I drink while she lists the manias that have plagued some of the great writers of modernism. Kafka's dendromania (forests). Beckett's agromania (open spaces). Lispector's crystallomania (crystals).

I make my way home as if there weren't a single light left on in the whole world. I dream I am a celestial bird with an enormous beak. I fall off my heavenly perch and, without meaning to, puncture the globe. I squark and flap while it deflates. All souls depart on the wings of the globe's rancid air.

The next morning, the window looks down on me. A white eye of light. Soon I am standing in the bathroom. In the lounge. Standing in front of a wall. I do not dare disturb myself. Now at the computer. Look calm. Look peaceful. Should an inchoate mob of creatures surge towards you, try to recognise them as beasts of your own imagination.

Strong coffee and meetings. Someone is taking minutes. There is a vote. It rains and rains.

My mind wanders off to a corner of itself where the same scene plays over and over—a man is hitting a pig, a pig is crouching, a pig is squealing, a pig runs the only way she can, towards the gas chamber. And on and on.

...

Earth. Trunk. Leaf. Moss. We can see none of it while we're in the Enclosure. The room is narrow. We take our seats. A question is thrown into the centre of the room, *How can we name unnameable things?* My head begins to unpack itself. Things like liquifying turd islands. The boldness of rats. Endless light or endless night. The force of a killer blow. Ribbons of entrails. Infectious miasmas. The whimper of a wounded animal. Like all these things multiplied over the thousands of moments in which just so many things take place in the world.

As we think, as we wait for someone to respond to this question—*How can we name unnameable things?*—I am also thinking of intensity and duration in a situation of pain. How curious that pain makes use of sonic lexicons with words like *frequency* and *volume*. That pain can be loud but, often, wordless. Or, if not wordless, then called on while still in formation. A new language, one specific to *that* body in *that* situation of pain. But also like a bridge, capable of carrying heavy loads towards other bodies, other minds.

This question about unnameable things hangs in the air between me and my students. Like a piece of old brown fruit. Who's gonna pick it? I look down, shuffle my papers just as if arranging cups. Now I am thinking about invisibility. A politics of disappearances. That animals worry invisibly, suffer invisibly and die almost invisibly. That power is vectored through this invisibility. That many people ignore the pain of the animals right in front of them. That this invisibility is aided by the cellular killing system of the abattoir, with its management principle of simplification and individualisation. That each worker is only ever given a slice of the killing. I look around to see who will name a single unnameable thing and so the second hour stretches on—a thousand potholed acres.

...

Instructions for a dying class: Hand out single sheets of paper. Place paper and pen on your desk. Place hands flat on the desk beside the paper. Close your eyes. Take three deep breaths. Feel the air, fully. See your own body as if from the very inside, fully. Now see it from the outside. Now ask a question that you would like this writing session to answer. Like, how are you flesh and not meat?

The aquarium glow of laptops. I am giving a talk on multispecies literary justice when a student conspicuously looks at her watch. A lecture can be one of two things: a hole dug to infinity or a pile of bones in need of flesh. I glance at the Head Tutor, she smiles. Does not recognise I am drowning, not waving. Afterwards, I hide in the toilets. Close my eyes. Cradle my skull. Overhear talk of *Moby Dick*—vengeance, instinct, fear, death.

Outside a bewildering number of people are rushing to learn the very next thing. People brush past. Mice, mice, everywhere.

...

Today everyone agrees lambs are five hundred to one thousand times cuter than all the other farmed animals. Students are debating this because Nacho, the bobby calf rescued from a dairy farm by a TV star, has gone viral. *I legit started crying when that baby was carried out*, says Human.

Then Rat proposed a morbid game, to write the end of the world in the fewest possible words. Everyone thinks this is cool, so I set a timer, we bury our heads then read our endings out:

> *A curtain fluttered.*

> *The Lord said.*

> *A blue feather and a woosh.*

*Goodbye!*

*The Queen ascends.*

These are intimate ends. Particular and uneventful compared to any traditional sense of narrative climax. They make me think of what T.S. Eliot wrote about the end of the world. That it happens, *Not with a bang but a whimper.*

...

I'm hiding in the bushes drinking coffee when a possum bursts out of a bin and looks around sharply. I'm reminded of the possum's innerness.

...

Twenty minutes into class and our topic is extinction narratives. We trip through the readings, turn over enormous questions. Our conversation about representation and multispecies loss develops a number of ripe and tender bruises.

I put this to the class: *What are the urgencies we need to write up to?* Silence hits down on us like a wave.

...

Octopus announced that she stopped reading stories about animals as a child. The mood in the room is that others did the same. The conversation rolls and darts, any discussion of animals and literature is swiftly and collectively shut down, as if through some telepathic radio signal I had not picked up on. Talk wings off in another direction.

Flocking birds dance in highly organised clusters like this to hunt, but also as a form of protection against predators. Have these writers imagined they no longer tell themselves, or absorb, stories about animals? Have they considered how much literature is founded on the assumption of human supremacy?

I want to ask these questions but neither the fullness nor the spectre of more-than-human animals can be brought into the room. This unvoiced prohibition made trivial feels hugely consequential. There are coffee sips. Silence. No one is even clearing their throat. I think, for a moment, that no one is even breathing.

My head is a deserted nest until a line from Jenny Offill's novel *Weather* falls in, *First, they came for the coral, but I did not say anything because I was not a coral . . .*

Later, to a colleague, I ask how we would proceed if we were to look into literature that carried its readers to the very edge of the slaughter pit. *What if we all opened a door to our brains and let the horrors flow in?* Turns out she's running late for a meeting.

. . .

*Drink, drink, drink, drink, drink.* In the old quad people chant this chant of infinite sorrow, which I think is a way of handling the world through a public ritual of smacking sorrow sideways. Or, through a ritual of blind obedience. Which is another way of handling the world. Of permitting yourself, your sorrows, to being ritually slapped sideways.

...

Some guy writing a thesis on Melville or Wallace Stevens, can't remember, stops me in the hallway as I'm legging it to class. He says he's got this joke. It goes, *Last year a student at the art school wore a GO VEGAN t-shirt for two weeks straight as research for his thesis. On the first day he was spat on, on the fifth day he was punched and on the tenth day a glass was thrown at him. Day twelve comes along and the student's father goes to meet the boy's supervisors. What's gonna happen, asks the father, when the kid actually leaves the house?*

In the shitty hall, with its smells of anxiety and burnt cheese, I stand before this guy's ugliness and in my own ugliness I smile, turn and bounce away. Deep in my guts a tree bursts into flames.

...

In class I imagine reading aloud a passage from a book that I had not put on the reading list. The passage describes the *live hang*—a process essential to maintaining the rapidity of chicken slaughter. The book is on the table in front of me. It would be an ambush, a kind of guerrilla pedagogy, words erupting from under a bush or behind a nearby building.

Words as improvised explosive device. The point would not be to explode, tranquillise or knock anyone out, but to stimulate: to speed up messages travelling between the slaughterhouse and the bodies of these few writers sitting in our Enclosure.

Class ends. I haven't read the passage out and I leave feeling stormy, thinking of Audre Lorde's words, *Your silence will not protect you.*

Lorde spoke this truth in her address at the Modern Language Association's Lesbians and Literature Panel in December 1977. In this speech, she acknowledged that writers who are trying to transform silence into language and action would need to scrutinise *the truth of what we speak and the truth of that language by which we speak it.* To do this might mean risking judgement, harassment, censure or contempt. Still, Lorde knew this work must be done by someone, by many someones.

By the time a two-hour tutorial is finished approximately two thousand broiler chickens will have been hung, dragged, electrified, neck-slit and boiled in a single large-scale abattoir in Australia. I sit alone with this fact and I'm filled with the sense of touching an infinity of dead others. I get the jolt of an urge to build altars of pure and tangible shock, of unending absence, of meaninglessness itself, of sadness and all that holds it back. As if this will restore the broken order of things.

...

Wet nose morning. I wake, try to recall my dream. A sea of young men screaming. A cow is being beaten with a stick.

One student writes about fire. Another about flood. The third about a scientist who sets out to measure his own soul. The questions line up. One wants to talk scene. Another dialogue. The third, character. In my head, *Oops! Page not found.*

# A Thanatography

There has been an explosion in the engine room of a cargo ship. Sixty-seven thousand sheep are left burning or drowning. They will not become earth, but sea.

Another ship spends two months at sea. Finds nowhere to dock. The animals onboard are thought to be carrying scabby mouth, a highly contagious viral disease of sheep, goats and occasionally humans. They are floating in limbo as the rope of life breaks loose. Six thousand die onboard.

A vessel capsizes and sinks while berthed in Brazil. The water moves. The wind shakes. Here is death. Five thousand drown.

Six thousand drown in a typhoon. Each one, like running in a dream, tries but cannot move. The havoc of a ship. Eyes sliding everywhere, gripping only onto blue.

More than eighty-five thousand onboard. Salmonella, *shy feeding*, heat. In sickness they find death. Five thousand, eight hundred of them.

One ship capsizes off the Black Sea coast. Almost fourteen thousand drown. How terrifying to be plunged into water.

To see the sea flicker all lit up blue. The strangeness of a blue death. To be tied, suddenly, to the pull of the sea.

Another ship, a notorious ship, is loaded with forty-five thousand sheep and a few camels. It sets sail. Soon it takes on another thirty thousand sheep and some cows. This is one of the few two-tier ships, which allow for animals to be packed into two levels of pens on each deck. This means the ship lacks any real ventilation.

Ventilation onboard such a ship, at any rate, is called *forced ventilation*, which means the temperature outside the ship, say forty degrees, is the temperature at which the ship is, so-called, *ventilated*. Sheep on deck might reasonably ask if the wind's been put to sleep. This ship sails for twenty-one days. Those unable to reach food and water, unable to lie down, unable to regulate their body temperature collapse in excrement. One and then another and another and another slam down, on their knees. The hot shock of falling on your knees; one and then another; of taking three hundred breaths per minute; of foaming at the mouth. Four thousand die this way. The very air puts a stop to their breathing.

Years later the same ship reloads in forty-four-degree heat. I try to picture the sheep. Staring at the pull of the sea. In a daze of heat they are boarded. Their lives unfinished.

To dream a dream of long-stemmed grass but die in a puddle of your own shit. It goes on and on. Seventy-four thousand onboard. Two thousand, three hundred go into death like boats go into the foam of a wave. Sixty-three thousand onboard. Two thousand die together in a burst.

Then, seventeen thousand die. Twelve hundred are unaccounted for. The company lost count of the dead and dying.

Bodies disintegrated in the heat. It *grew difficult* to track the dead. A heat curse on them all. Did they have even a moment to set their faces to the wind? Lift their faces to the air to ask: *Who is there? Who is there to help us?*

A typhoon. Forty sailors and 5,800 cows go missing, feared drowned, in the East China Sea. Folded into the silver-grey of battering waves. They lift, they swell, they bloat. I close my eyes and try to imagine as each one sinks into their own horror.

Two ships set sail. One carries nine hundred calves. The other 1,800 young bulls. Both arrive at their destination late December. Both are refused entry. Bluetongue disease has been detected in the province of their provenance. Maybe these animals are carrying the disease, too. Clinical signs of bluetongue include: fever, lameness, nasal discharge, haemorrhaging, swelling of the lips, tongue and head. Sometimes, a swollen bluish tongue is detected. The calves are sent from port to port. They spend days without food. One ship is at sea for two months, then sent back to its origin. Vets declare those onboard unfit for further travel. They are slaughtered. Each birth has only been a death warrant.

For three months the other ship rolls, heaves and pitches. Food shortages, water shortages, lack of movement, lack of air. A build-up of shit. The dead are chopped up and thrown overboard. In the headlines their odyssey is called a *Cruise from Hell.*

Ancient Egyptians theorised that the journey to death happened in several stages. The first was encompassed in the vehicle of transport. The choice of route was not up to the passengers; it depended on their status.

# Calculus

What if the concept of survival was simply not for you? That survival's equations, its calculus of infinitesimals, did not include your number? I have been searching online for a Beginners Guide to Survival. Though I haven't found one, I have learned that the word calculus originally meant small pebble, because pebbles were used in ancient times for counting or measuring distance travelled. The travel of variation, the travel of continuous change, the travel of propositions, the travel of process. These might be descriptions for survival, which is a form of travel. Which is perhaps why we ask each other, *How are you going?* Which is to ask, *How are you surviving?* Which is to ask, *How are you coming through, rebelling against or living longer than the things that might kill you?*

How many acts of injury can a body withstand? To walk without shoes? Find no grain? To be isolated, truly isolated? To be fed poison? To risk drowning in your own shit?

Is there an anatomy of survival? A chart of its most important or intimate parts? A way to learn the tender architecture of survival's tissues? To fix what is broken or extract its many tumours?

I've been looking for a guidebook called Method of Survival Theorems, or something like it. Not a preppers manual about food caches, knot tying, dead bodies, firearms and refusing to fall in line with the unprepared masses. A dream book that lays out different kinds of survival as concept and reality with chapters on the relations between survival and self-reliance, survival and production or reproduction, survival and care, survival and poetry. It would have chapters on the everyday, the urban, the industrial, the wild, the vanishing wild, the ancient, the repetitive, the torturous, space, lack of space, noise, endless noise, the production of death, the production of labour, of labouring bodies, deadly bodies, sick bodies, killable bodies. Chapter headings like Getting Home, Pure Violence, A Temporary Event, Heroic Refusals, An Exercise in Chance, Permanent Change, Imminent Catastrophe, Liberated Zones, Stupidity, Vulnerability and Ruptures of the Everyday.

I drift along my bookshelf and cruise the web searching for this guide that does not exist wondering exactly how ugly is my desire to have such complexity spelled out in the furred ink of a book I can put beside my bed and abandon.

In the stories of my childhood, abandonment was ever present. That's how I remember it. Women are abandoned by heroic men in Greek myths. The women who have stayed with me are the ones who became suicidal or murderous, like Dido and Medea. Medea who said, *Of all creatures that can feel and think, we women are the worst treated things alive.* Had anyone asked her, *How are you going?* Or, *What's up?* Or ventured further to make a genuine inquiry into her wellbeing after she was tricked into love and marriage, had gone on a killing spree and been abandoned with two young kids while living in exile?

In fairytales, children are left wandering deep in woods by moonlight, or else they're locked in towers. Or poisoned and left for dead. Or turned over to hard service by parents who can no longer feed or care for them, or never wanted to. Sometimes the parents are peasants, labourers, pushed to extremes during long, cold famine. In these stories winter is the season of abandonment. It is the season in which short days feel so long and things transform. Like breath turns to ice. Labouring bodies become broken bodies become hungry souls. And bodies that shake, at first, with cold begin to shake with hunger. And hunger that was once felt as a hollow or expansive hole in the body grows sharp claws, mangy fur, a bloody tail, which is an image I recall from Aki Ollikainen's novella *White Hunger*. Which is about a devastating famine in Finland that lasted from 1866 to 1868 and came after years of small crops, then a cold summer, early frosts, then storms, heavy rain, flooding, then more winter, endless winter, that destroyed the crops. The novella follows a woman, Marja, a farmer's wife, who walks with her two young children towards St Petersburg in search of bread and it asks, *What does it take to survive?*

What does it take to survive is, perhaps, the question on which stories of abandonment concentrate most intensely, ask most openly. These stories are records, in writing, that stand in the place of narratives that have fallen out of knowledge—that are lost, have faded, can be known only by proxy. Narratives about people who walk into places and times from which there is no return, on journeys without companion, or without companions who survive.

Another way to talk of survival is as a process of keeping the body and the soul together, which is another way to talk about having enough food on your table and money in your pocket to stay in one piece. Otherwise, the body starves, the soul can grow sick.

I do not think anyone asks Marja *How are you going?* Instead, they say things like *Jesus, Marja ... Jesus ... help.* Or, no point. Or, this is all I can do. Or, your faith is being tested. Or, put up a fight. Or, I cannot give you more. Or, forget them. Or, how will you ever make it? Or, simply, frozen.

# The Jungle

An academic I once met worked undercover at a slaughterhouse in Nebraska for a year. He observed how cows vomit after they've been stunned, *depositing a rank greenish substance onto the floor that mixes with the blood flowing from its head wound.* He noticed how heads and necks will tremble as if they are experiencing a seizure, how heads can fall, necks stiffen, tongues hang limply from mouths. If, however, the angle and strength of the bolt isn't enough, a cow can remain conscious, bleeding profusely, thrashing around while the stunner tries to bolt them again. In the case of bolt stunning, when cows kick vigorously, even after their spine has been severed, it's because the circuits that provide the rhythmic movements for walking are located in the middle of the spine. In this situation, the readable body becomes unreadable and then it is reassembled into a new text.

What words are there for the hours at slaughter? For the stench, the cold, the heavy loads, the life unworthy of life? The face of each cow is punctured like a piece of old fruit. I try to imagine the unnamed, first of the morning, spraying a message on the floor in urine to those who would come after—
*Don't come here!*

To stun a cow, the gun must be placed perpendicular to the skull at an imagined centre point—the X. A hiss, a bang. I imagine the way a cow dies mid-air. The first, then second, third, fourth, fifth. Like leaves they will be fallen. Though the captive bolt gun used in abattoirs has the same effect as a firearm with a live bullet, the rapid insertion of a bolt through a skull is said not to hurt.

Slaughtering might consist of a sensorium of horrific sounds, tastes, acts and smells, but it is a job, one of the most significant industries in Australia. Australia is, for the moment, one of the most energetic meat-eating nations in the world. Since the 1960s Australians have consumed approximately 110 kilograms of meat per person per year.

The carotid artery and the jugular vein must be cut while the cow is suspended upside down. For hygiene, the cutting is to be done with two knives. One to get through the skin. The other is for the veins. In those last moments, lives are turned towards the underworld. And they call it casual work.

In Upton Sinclair's novel *The Jungle*, which is about Chicago's Union Stock Yard, the question *How are you going?* is transformed into *How did we get here?* Everyone in the world of that book is poor or sick or both and tottering on the brink of ruin, which is the world on which capitalism stands, which is the world in which the novel stands, too. A world where work is endless and money is fugitive.

It is a novel about life and death in the vast world of Packingtown, part of the stockyards. If, at first, a visitor is lost to the wonder of machinery, the vast pumping networks of muscle and steel, the brutal speed of the industrial assembly line, the logic of pork-making, one soon becomes disoriented by the streams of animals a river of death that flows from the pens outside to the killing beds inside; by the squeals and grunts that grow louder and more agonised as pigs are hung upside down, as their throats are slit, as the life-blood rushes over their faces; by the human nature of their protests.

In these ruins, bodies and souls are worked to a pulp. Some souls scrape hairs off the dead. Others deal with the entrails. In the course of it all, fingers are lost, workers fall into cooking vats, or get blood poisoning from canning, or rheumatism from long days spent in cold rooms. Fountains of blood splash onto the floors. This work shows a truth about labouring bodies, sickening bodies, killing bodies and killable bodies. The truth is that survival in concept and reality is not configured to include everyone, not by far. *The Jungle* shows that survival in the slaughterhouse is slim and its opposite enormous.

When he wrote *The Jungle*, Sinclair wanted to blow the lid off the rotten, diseased and dangerous world of Packingtown. A world in which tubercular steers and pigs died of cholera in the box cars of trains that carried them to Packingtown and were sold for meat and lard anyway. Packingtown *was a place where men welcomed tuberculosis in the cows because it made them fatten more quickly.* And there were cows who had been fed on *whisky-malt*, the refuse of the breweries and were, as a result, covered with boils. The factories were places for a Dante or a Zola, Sinclair said. Places where violence reached an irrefutable peak, and then went farther.

Hour after hour, day after day, women twist sausage links and they do this until the day they are buried. Sometimes a steer breaks loose and, slipping on blood, runs for freedom.

On freedom. Though one is free to work at breakneck speed, one is not free to live because death is the source and fountain of all life in Packingtown. Animals and people crawl about and weep and this is called Life. And what is worse than the crawling and weeping of Life is called Survival. Mere Survival. The haul of Survival. And it would have to be all. Until the possibility of even that is lost in the melee.

# Dictionary of the Vulgar Tongue

In Jean-Baptiste del Amo's novel *Animalia*, The Beast is a boar who doesn't die though his mother is shot in the head five times and the last word she hears is *bitch*. Bitch is an angry dog. Or worse. She's a dirty angry dog and a whiney, complaining woman.

Bitch has travelled to this moment all the way from the year 1400 like a storm roaming the sky. Like a doggess. She is good on the road, walks well, has outwalked generations.

Bitch from the 1930s means 'to spoil'. Somehow related to *bicched* from the Middle English, meaning cursed or bad, as in *This fruyt cometh of the bicched bones.* Is there an animal alive on Earth right now who does not feel themselves to have cursed bones?

Picture the word bitch as it falls through the hole punched into the sow's interior. Imagine leaping—fully grown and fully armed—into the house of the sow's body. Go skull wise. As you leap, recall that the root of house, shelter or refuge is *hos*, which, Elaine Scarry writes, moves out along two paths. One in the direction of a protective space—host, hostel, hospitable and hospital. While the other heads off to a world of sacrifice, of sacrificial victims—hostis, hostility, hostage.

In the English translation of the French novel *Pig Tales* by Marie Darrieussecq, the word bitch appears only three times, as *Son of a bitch, a real bitch* and *the bitch should be shot.* Despite the infrequent use of the word in the English version, the original French title, *Truismes*, forms a constellation of associations that pattern onto the meanings of the words bitch and whore. The protagonist is shaped by her supporting cast of characters as a spoiled, complaining, she-dog of a woman, though the point of the novel is that she metamorphoses into a sow.

A truism is a statement that is so obviously true it says nothing new or interesting. But in French the word does a lot more work. *Truismes* offers graphic and phonetic connections to the phrase *grosse truie,* fat pig or sow, and, I learn, is a common insult directed against women. It tramps from the biblical to the pornographic for it also means *To reap what you sow,* unclean woman, seeding and insemination. *Truie.* Female. Noun. Sow. Belongs to *cochon.* Masc. Noun. Pig. Also meaning dirty, naughty, indecent, filthy, smutty, slutty or to soil.

This is sowography—the writing of pigs. By which I mean what human culture makes of pig culture in language. By which I mean we learn nothing about pigs through the study of sowography. Only that language is turned into an elaborate and inescapable weapon used against them.

# Snuff

How to write *of* death, or even *from* death? How close can writing come to death? Can one measure the nearness of biography—the writing of a life—to thanatography—an accounting of the dead? Is it possible to make a form of writing out of the shape of an afterlife that might be without shape?

Would its principle, formless forms, be ephemeral lists of soul guides, metal tools, stones, pottery or modes of transport that carry the dead? Or would they be poems made only of the last lines of unsung songs? Or half thoughts? Or parataxical monologues, in which unlikely things are arranged side by side, like Beckett's *Not I: out ... into this world ... this world ... tiny little thing ... before its time ...* Or would we need dictionaries of unspeakable words? Or dictionaries filled not with words but with radiance? Or dictionaries filled only with excessive space for grieving or thinking or remembering or forgetting?

I think you would need death sentences. The grammar of death. The syntax of passing out. Of falling. Of sending forth. Of gliding away. Of shooting one's star. Of being shattered or scattered or *completely* scattered. The lexicon of death. *That* might give death its fullest expression.

When I think of writing the dead I think there is so little to say with precision because so little can be copied from that passage beyond. The passage that moves off into a landscape of textures called Nothing. A path littered with things known only as The End. The End. A location, a zone, a road, a denuded road. Say you stand by that denuded road and watch trucks ferrying the dead. Say you copy down everything that you see, hear, smell and feel. Still, you would not have written the dead, only the procession, only the entrance, only the pathway to, but not the thing itself.

*The act of writing, or talking about one's death,* says Edwidge Danticat, *makes one an active participant in one's life.* But there are no monuments for the dead at factory farms or abattoirs. No obituaries. No eulogies. No Facebook accounts or virtual cemeteries, in which they might receive messages to read in the hereafter. There is only flesh, bone, blood, skin, brains, vomit, fear, sweat, spit, shouts and cries. There is only cutting throats, hanging livers, peeling skins, extracting arseholes and emptying stomachs.

...

A cow arrives. Is moved off the truck. Is moved to the driver (who uses *electric prods, peddles, whips, and voice to drive cattle*), is moved to the knocker (who uses an *air gun to drive a captive-steel bolt into the foreheads of cattle*), is moved to the sticker (who *uses hand knife to cut jugular veins and carotid arteries of cow*), is moved onto the ear cutter and nose cutter (who *uses hand knife to slice off right ear and right nostril of cow*) and then is moved on and on. Confused and endlessly renewed, animals pass through to carcass lines, headlines, offal lines, liver lines and finally to coolers. Just like time, they are killed in a measurable passage. And, just like time, the process will not be reversed.

. . .

Soon I will write about the shape of disappearance because I cannot, otherwise, tell how disappearance works. I will write about sinking and negativities. Of incinerated mornings. The arts of hiding. Pictures of nothing. Of anti-display. Or, the unlikely parsimony of display. I will conjure, with nothing but the whole existence of a word, the curve of the shoulder of the bereaved because it, bodily, speaks the exact slant of loss.

I go searching for disappearance just as if it were a river, blank and undulant. A river of last light. A river of exits. A river that is bigger now than it has ever been. That swells up, like a puffed face, and smells rotten. I tell myself, not only will I find this river, I will follow it all the way to its source.

Writers lament the disappearance of words and languages that are unravelled by species extinction and alterations of place. As land and water are changed, one forgets what was there before unless we have language and stories that remind us. Ecologists call this forgetting the *shifting baseline syndrome*. Ruined landscapes become the new reality. In 2007 the Oxford junior dictionary removed words about nature—blackberry, moss, acorn, and bluebell. In 2012 cauliflower and clover were axed. By 2015 prominent nature writers and novelists had sent an open letter to Oxford University Press. They protested that stripping the dictionary of words relating to nature threatened to further unravel the link between society, culture and the natural environment.

Those who write of commodified beings have this problem in reverse. We begin with a ruined landscape of language. Beyond statistics, herd counts, units of value, weight, profit, loss and liquidation we have only the song of nothing. Despite the population of farmed animals, and the sheer profusion of their bodily products, literature that attends to their lives and deaths remains marginal. The slaughterhouse has not been turned into a poetic analogy for the depths of the human unconscious. Why?

...

Three a.m. I'm online flicking through a photo essay documenting the inside of the Danish Crown slaughterhouse in Horsens, the world's largest exporter of 'pork'. It is celebrated as the most modern slaughterhouse in the world. One hundred thousand pigs are killed here each week. And you can watch it all online, in person. What do people hope to experience by watching animals get slaughtered? Connection? Cleansing? *Schadenfreude, joie maligne, skadefryd*, pleasure in another's suffering?

Slaughter tourism—a niche of a tourist niche—intersects with dark and disaster tourism, histories of public executions and the thanatological gaze. All of which makes me think of the word *snuff* and that genre of film that is supposed to have begun, at least in concept, in 1907 when the Polish-French writer Guillaume Apollinaire wrote a story about photo journalists who stage and film a murder because, even then, crime news got good ratings. I think of the bobby calf shooting films as a kind of low-budget horror and propose to myself a new genre of film—*abattoir noir*—which might be a film style or might be a form of reading, I'm not quite sure. Though if it were the latter, I'd treat it as a kind of cultural detective work. A minoritarian or marginal interpretation of narrative artefacts not usually seen as crime narratives.

. . .

The origin story of Blood Pattern Analysis starts in 1895 when the first study was published at the Institute of Forensic Medicine by Dr Eduard Piotrowski titled: *Concerning the Origin, Shape, Direction and Distribution of the Bloodstains Following Head Wounds Caused by Blows.* At the Institute, in Krakow, Piotrowski sat a live rabbit in front of a paper wall and beat her to death with a hammer. An artist then recorded the patterns. Later, other rabbits were beaten with rocks and hatchets—all kinds of things—using varying positions and angles. The results all recorded.

. . .

Visibility has never been as easy as the equation, looking at *x* means seeing *x*, means *x* has been seen. There are poets for whom disappearance is a felt presence. Poets who enunciate the world that has been stashed away, who utter appearance by way of elimination, who write that there is no friend as far as the eye can see. Lately, I have been grabbing onto poets who write against certain evaporations. Like Wisconsin poet Kimberly Blaeser, whose *Poem on Disappearance* instructs, *Draw nothing around a crumbled bird body—no wings.*

When I was six, maybe seven, my grandmother would feed me corned tongue sandwiches, an old-world Jewish comfort food. The tongue is traditionally preserved by pickling in salt brine, garlic, pepper and spices for several days. She would buy them from a deli in Caulfield, place one in front of me—a tongue in white bread with butter and pickles. I didn't recoil when the organ was named. I sat at the table that was covered in a white lace cloth. My heart was beating. When I peeled back the bread to look I felt only curiosity. I used to lick the tongue before patting the top layer of bread back down and eating. I looked at the tongue and all I saw was food. That was my first experience of disappearance. The tongue was not itself. It had been neatened. Processed away from its origins. Cooked in water for hours. Whole in itself, only bodyless. Combusted, levitated, expanded, transformed. A banal magic. A disappearing act.

I google transfiguration, untransfiguration and conjuration but can find only spells to vanish certain animals, not call them back. The vanishing spell, I learn, is the most complex of all. Snails are relatively easy to vanish because they are invertebrates. Mice and cats, being mammals, are much harder. What makes a snail so easy to vanish? Not the majesty of their gentle heads.

Abattoirs themselves have done a disappearing act. The word from the French *abbattre*, 'to slaughter', appeared at the beginning of the nineteenth century with the creation of the first public slaughterhouse in France. This, in turn, sparked the movement of animal slaughter from public and civic to private and guarded spaces.

In Melbourne, public slaughtering was shifted away from public gaze in 1849 when it came to be quarantined below what was called Batman Hill on Wurundjeri country. In 1861 civic abattoirs were shifted again to Flemington, where they came to perch on the Saltwater River, the Maribyrnong.

Municipal abattoirs were also set up in an arc stretching from the Yarra River in Collingwood to Williamstown. Along that arc, around three hundred tons of blood was pumped into the rivers and the bay, offal too. Slaughterhouses occupied low-lying sites along waterways. The sites were badly surfaced, badly drained, poorly ventilated. Blood and offal accumulated freely. Later, slaughtering facilities were sent to run aground in outer urban communities like Laverton, Brooklyn and Albion to the west, Dandenong, Cranbourne and Pakenham if you go southeast from the city centre. As Gomeroi poet Alison Whittaker has written, slaughterhouses have *been dutifully turned away from polite society*, but even so, *they are hardly invisible*.

This creep from open to closed catalysed the transfiguration of industrial spaces of animal extermination into a dark cinema of the human mind. *Abattoir amnestia*—a forgotten place of wrong. Not a secret, but a disregarded depth.

# O Blood as it Pours Down the Drain

In the archives I expose myself to the singular topic of slaughter, to explore the textures of abattoirs as culturally disregarded depths. I work through folders of images, wanting to assemble a visual cache relating to livestock industries in southeast Australia from the time of early colonisation through to the 1980s—animals, landscapes, architectures, implements and people in various scenarios of isolation, abuse, incarceration, surveillance and death. There is a bloodless torso, a bull at auction, merino sheep onboard an aeroplane (circa 1950), a horse freshly killed by a knacker.

O guns. O knives. O squeeze pens. O whips. O paddles. O shackles. O hydraulic knife. O air knife. O hooks. O puller. O clamps. O trapdoor. O rips. O holes. O incisions.

There are charred bodies beside a fence, hands covered in blood, a horse pulled onto a train. I assign aliases to the unnamed people in the photographs: Victor 'Eggs' Benedict, Jason 'Pork' Chop, Robert 'Milky' White, Chuck 'The Steak' Taylor. Hunched over the lightbox, the space-time of long-haul truck carriages and trailers stacked with sheep end to end under a hard sun is illuminated. The sheep will be driven to what is, for them, the end of the earth. That's all. A highway becomes a river separating the living from the dead. It is also, always, just a road.

I try to imagine breaking through the eternal seal of death. *Any fool can get into an ocean*, says the poet Jack Spicer, *But it takes a Goddess / To get out of one.*

. . .

Today, thinking about public executions, decapitations, punishment as public display, the head as a trophy, wall-mounted, stolen to make the deceased difficult to identify, unravelling the singularity that properly belongs to a face.

. . .

The face of a cow is loaded with questions of power, vulnerability and violence and in their eyes is the look of one ravaged by war. Like the story of Perseus, who can only approach Medusa with the use of Athena's shield as a mirror, between the public and the slaughterhouse there are significant mediating apparatuses that clip, angle and bend our vision, so we can drive past factory farms without seeing their disfiguring violence. We can avoid the *phrisso*—bristling, shuddering, stiffening—that could lead to petrification.

Is it possible to develop a *philia*—a certain kind of imaginative desire—to counter this *phrisso*? And how to stop this *philia* from desiring violence?

In the archives, again. This time the images translate themselves into bodily sensations more so than thoughts. Nausea
and thirst. A ringing in my ears. Stiff neck. The end of a world
in every image.

· · ·

I can't see the facial expression because the camera angle is
pointed straight down, a god's-eye view. The lighting is stylish,
shadowy. It's impossible to tell the time of day. This photo was
taken in the 1990s. Part of a series that shows the interiors,
employees and work practices of Hardwick's Meatworks
at Kings Court in Kyneton, a place that lies an hour outside
of Melbourne if you head straight up the Calder. The image
conveys a distinctly noir aesthetic with its high-contrast mise
en scène. Starkly lit, the white head is a sculptural form. Not
a being or an individual, but a partial or dream object. There
are two boxes of bolts on the shelf. One is part used, the top
ripped clean off. This makes me think it won't be put away at
the end of the day. It will be emptied. Beside the shelf hovers
the cow's head, still intact. No sign of a shooting. I have to
think that her eyes are not yet sightless.

The image does not just undo the cow's figural unity; it takes
away her life. The more I look, the more it brings to mind
stories of other severed heads, like Medusa, those who enflesh
an apotropaic function. That is, give body to a warning.

· · ·

Darkness can come after a flood of photons. Dark volumes and partial blindness are produced from too much light. This darkness is different from shadow. It is more like an eclipse riddle, the theory of which states that from a certain vantage point bodies can appear to disappear but only because of an optics of nearness or farness. Though philosophers of shadows debate the point, some say that perception favours what is far.

...

Dark tourism dates back to the Middle Ages with the thanatopic tradition—a contemplation of death. There are different shades of dark tourism, a spectrum of supply ranging from darkest to lightest. Identified and scaled according to the intensity of interest on the part of the tourist, the degree to which that interest is exploited for commercial purposes and its spatial, temporal, ideological and political factors. If the darkest tourism amplifies the experience of death by the recentness of tragedy, the abattoir markets a trip to intensity, a holiday in hell.

...

Last night I dreamt of zombies. At first these dreams came with attitudes of resignation. They quickly took on the shape and force of metaphysical hymns sung in bellicose voices. They sung the principles of organic vitality, of bodies changing, they expressed laws of bodily cuts and holes that are vicious and unclean. I was led to understand each hole held mysterious properties, was elevated above the ordinary. When I woke my mind was a broken bowl.

. . .

The darkest part of a shadow is called the occlusion shadow because it can't be reached by reflected light. Technically, you could say all shadows are kinds of occlusion. But I am told by a physicist to reserve this particular term for the dark space directly beneath an object.

# Joint Labour of Generations

Why is there no great novel on the pain of horses kept in 'pee barns' for the production of the hormone replacement therapy Premarin—a drug named after its key ingredient, PREgnant MARe's urINe? The horses are fitted with a rope or chain harness and a urine bag, strung up virtually immobile. Urine is collected from October to March. Foals are born shortly after collection and the mares are re-bred within a few weeks of foaling. It is uncertain whether the practice of depriving mares of water to concentrate their urine still happens. The foals will be auctioned off by September. At a certain farm in Canada when mares resisted separation from their babies they were whipped, kicked or beaten with an electric prod until they could not stop their child being taken away. Male foals would be auctioned for slaughter, females were kept to replace their worn-out mothers. What would those mares rather be doing? What would they rather their children be doing? How many clocks have skipped a beat in grief for them?

Plato wrote of the body as a prison, a tomb or coat for the soul (which he felt was the true self). He located the uniqueness of a person within the upper torso and the head. The section of the human body between the midriff and the navel was considered 'bestial'. He wrote that land animals came from men who were so stupid that they had ceased to use the circles or revolutions in their heads (reason) and, instead, followed their spirits. Because of this, their heads were drawn towards the earth, forcing their arms to flop down in front of them and act as a kind of support. Thus four-legged creatures were born. Following this logic, he wrote that men who were stupider again were given even more legs. God, in his benevolence, handed out six and eight to insects and spiders. Ranking lower in the hierarchy of wits are those who lay upon the earth with no limbs at all. Thus certain reptiles and amphibians came into being. Finally, as punishment, the most ignorant of all men were sent to live in the depths of the water. Their souls tainted with transgressions, the gods believed they did not deserve pure air but *shoved them into water to breathe its murky depths.* The depths signalled nothing less than the depths of their dull-headedness.

Animals living in service of man were the only ones who belonged in Plato's vision of an ideal country. Aristotle developed this idea, wrote that animals existed only for the sake of man. Later, the Roman philosopher Cicero said that everything on earth was created in service of something else: corn and fruit were made by the earth for animals, and animals were made for man. These same ideas appear in Genesis—*Let man have dominion over the fish of the sea, and over the fowl of the air, and over everything that moves upon the earth.* Augustine said that the sixth commandment, *Thou shalt not kill,* applied only to humans. Only beings associated with reason could be murdered and since animals were not associated with reason they technically couldn't be murdered. The medieval theologian Thomas Aquinas chimed in with his idea that the life of animals was preserved only for man. This cleave was wrenched wider by Descartes who said that animals did not, do not, feel pain.

This is philosophy, religion, science and narrative as ritual exclusion. Human is separated from animal. Animal is separated from soul. These are conceptual moves Val Plumwood calls *hyperseparation*. Or, the concretion of a licence to kill, a permit of abuse. The 'animal condition'—of life unworthy of life—and human supremacy continue to be produced through canonical and citational politics. Or, the joint labour of generations.

# Holiday in Hell

Can you imagine a blockbuster film about slaughterhouses? Like a war film. A holiday war film. Like the ones that get released at the cinema on Boxing Day. There's screeching, hysterics and big battles on the big screen. It'd be called something corny and shit like *Holiday in Hell*. In the beginning we'd see young men and women asking to be sent to the abattoir. They want to be at war. Not a distant war, a local war. A war of *resources*. *This* kind of war. They imagine leading cows and pigs up to the stunner's box, stoic as a hero. A duty. They imagine nerves and then relief. What happens is different.

**Cast**

Man

Co-workers (chorus)

Doctor

Endless cows (chorus)

## Act 1

Sunlight grips the man's face. His body is suspended on a mattress. He sits up. Says to camera, *You have to put them out of their misery.* One worker swears like a trooper and calls himself The Surgeon. Another sits against the wall in the tearoom and routinely bursts into tears. There are jokes. Laughter. Basically, in this world, there is meanness, sadness and friendship. Days pass, then months. They are told they are fulfilling their duty. One smiles all day beneath his hard hat. His job is to grasp livers. He works with his two hands. Pulls livers up and off sharp hooks. Dark purplish blood pours down his gloved hands. A trickle at first. Then spurts. Secretly, he starts to question who he is.

The Man dreams all night long. Crawls out of bed, 4 am. No animal accepts death meekly. He goes numb. Carries out his duties. Starts to picture his own death, small and gruesome. A brown cow collapses in the knocking box. That shuts down the whole production line. There are shouts down the radio. The cow is shot with a portable knocking gun and winched away. Soon another goes down. This time those coming in behind are made to run straight over the cows in front.

## Act 2

The Man says to the camera, *When you shoot a bolt into a skull's grey matter, there's a crack and a wet splat.* Killing someone can take a second or minutes. He pokes a cow with an electric prod, causes them to mount the cow in front. Another gives birth out in the pens. He looks directly at the animal who has just given birth out in the pens. She is not asking anything from him. She is just trying to pass her afterbirth. He waits, as per government regulations, for her to pass the afterbirth. Finally, he sends her in.

They strip everything off the dead. Ears, eyes, feet, lips, cheeks, skin and stomach. Even arseholes. Unborn foetuses are tossed onto a heap. Afterwards people ask, *Where did you do your duty?* He keeps his lips tight. Recalls the vomit.

### Act 3

He's living in this world, but beyond it. Alive, yes. But off in a different direction. He goes home and smokes. Tries not to drift off the edge. It's been years and he can't smell anything but the abattoir. Not flowers, tobacco, soil or perfume. Never in all his life has he wanted to smell a fucking flower so much as he does now.

Diagnosed with a mild nervous anxiety, a doctor gives The Man something to help. The idea is to help him return to the world. But he's cursed (no he's not). Just lives for a fairly long time and never forgets certain scenes. He looks in the mirror. He looks and looks. Appears unharmed on the surface.

# Raw/Cooked

I am on my way to Hell. Or I am already there.

> Bloodied floors and walls bring to mind words like scrape, brain and miscarriage and lead me to a place of remote loneliness, full of heat, speed, sweat and noise. A place of the dying and the freshly dead. A bloodied hand is washed. Skin is ripped from a body.

> Downstairs, we had been given a warning—*Viewer discretion is advised*. Though perhaps it should have said *ABANDON EVERY HOPE WHO ENTER HERE*.

In the small gallery space are sixteen photographs in which blood covers walls and bodies in finger lines, drips, splatters, spats and layered flicks.

> They remind me of abstract paintings that work to create new scales of attention, new definitions of surface and touch, like the work of British artist Jenny Saville, who layers paint so thickly onto canvas that each one takes on the dimension of molten, mutinous flesh.

In contrast, the watery red in these photographs turns things like walls, sinks, rubber smocks, hairnets, faces, arms and knives into a form of kinetic art—which relies on motion for its effects—though the images themselves are still.

I pass from photograph to photograph and perceive a new kind of syntax.

Text written in blood, saliva, sweat, fur and water. Writing that reveals a bond between language and action, pain and imagination. This new language arranges and rearranges itself in my head as I walk the room:

Here is the weird language of a bloodied hand. Of face without eyes. Of a black hole wetly edged with blood, hair and shit. Of viscera at the threshold of another worldy dimension.

. . .

Today we're supposed to be discussing Deborah Levy's experimental novella, *Diary of a Steak*.

It's a meaty rewriting of the history of hysteria, which Levy uses to make an ecopolitical critique of bovine spongiform encephalopathy, BSE, or mad cow disease.

The book is a lunatic report and a post-death narrative that reveals a porous relationship between the degenerated lives of cows on twentieth-century farms and the degenerated lives of women in the Pitié-Salpêtrière Hospital in Paris where, in the mid to late nineteenth century, Jean-Martin Charcot (the 'father of neurology') and later Sigmund Freud studied hysteria.

Buttercup, a steak infected with BSE, delivers a darkly humorous, polyphonic protest against intersecting violences of patriarchy, humanism and carnivorous nationalisms.

Do you remember BSE?

A thumbnail history: it occurred as a result of cows being fed, among other things, *mammalian proteins* or meat-and-bone meal, including brain matter and spinal cords of other slaughtered cows and sheep. This caused spongelike patterns to develop in the brain tissues of cows, which caused them to develop head tremors, nervousness, aggression and progressive incoordination.

1984. A farmer contacts a local private veterinarian to examine a cow called 133. 133 had an arched back, kept falling over, had lost weight. By early 1985, she was dead. Soon, five more cows died on the same farm. The veterinarian called the condition 'Stent Farm syndrome' after the farmer. Soon, another two cases were discovered on the farm. The first of these was shot in the head, her brain tissues sent for analysis. Unfortunately, there is no diagnostic value in a shot brain and so a second cow, 142, was sent, alive, for pathological testing. She was slaughtered and given over to a post-mortem examination. Her brain had turned spongy.

At the same time, on a farm in Kent, a Friesian/Holstein showed changes in her behaviour. She was aggressive and suddenly lacked coordination. Others experienced similar things. By December 1986 more and more cases had been identified on farms in Bristol and Wye. Pathologists were finding strange formations in the brain tissue of these cows. Over the next fifteen years a million cows would contract the disease, though 4.7 million were destroyed as a precautionary measure. If a single cow in a herd showed symptoms, all were *eliminated*.

It might have been in 1988 or 1996 when I saw news footage of cows burning, smouldering, lying in piles on the ground, their bloated bodies hanging mid-air, lifted by tractors.

This was my second experience of transmutation—a lesson in understanding that farms are killing fields, bodies can become biohazards, and tools are also weapons. Claude Lévi-Strauss says that over the millennia humans have so profoundly transformed animals bred for food from what they were into what they are, they can no longer be called animals *but nutrient laboratories*, in which the organic compounds necessary for human consumption are developed.

This form of transmutation has relied on the disappearance of beings through radical biological reorganisation. The creation of a new order of beings for the purpose of industrial-scale production. Aka: extinction by eugenics.

*Diary* was published in 1997, just one year after BSE was found to have jumped species from cows to humans who became infected with the new variant of Creutzfeldt-Jakob disease.

The book was almost entirely ignored at the time of publication. Pushed into a kind of marketplace quarantine. The single review it received was due to its cover art depicting Buttercup, the uncooked steak, waving a little white flag that pleads, *Do you want to hear my erotic music?*

The reviewer wrote that reading *Diary* on public transport in Britain at the end of the nineties, when anxiety about the disease had been lurking for a decade, felt like wielding a minor weapon.

In class, I ask if someone would like to begin reading an extract out loud.

Just a page. A stanza. Just a line. A little line. A lineette. I shuffle papers, look down at the desk. Try to think about what the students might be thinking about. Silence piles onto silence. I wonder if this is what a spring die-off would sound like, when no chicks hatch, no apple trees blossom and there are no bees to sing. A rebirth of silence.

What will we inherit from this silence, each of us? Silence that feels naked, that circles like pigeons under a darkening sky and falls, too plunging, between us?

The moment grows unimaginably long, becomes space for an emergence or a ruin. A retreat from speech or a refusal. I can't decide which, but no one will make eye contact with me.

How to distinguish between different kinds of silence: remembrance, stealth, protest, (self) censorship, guilt, obedience, complicity, support or the slack-jawed void of ignorance?

There is a figure from the world of carnival who I wish to carry with me and unveil in moments like this. She is the Dutch saint Aelwaer, the patron saint of quarrellers, rioters, troublemakers, revellers and musicians. She wears a screeching magpie on her head and holds a squealing pig beneath one arm and a wailing cat up in the other. She has been thought of as an anti-saint, the Virgin Mary in reverse. Riding upon an ass she is all uproar and I want to take her as my patron.

Yesterday, joke guy leapt out at me again:

*What do you call a visit to your local abattoir?*

Beat.

*A major moo killer.*

Tonight, I am at my computer.

The more I write the more I panic. Never before in the history of the world has there been such an example of mono-specific, vast bird biomass as there is on Earth, right now. The rate of chicken corpse accumulation is unprecedented in the natural world, what ever that means. Their shape, their chronic oddities, their anxious hearts have become our deranged geology.

Writing, if it is capable of doing anything at all, has to incite total animal liberation. Has to move society towards a new form of existence, or it will be a failure.

I would climb inaccessible mountains, move through dark forests, write only in states of ritual purity to do this work. Instead, I go wandering through the desolate wilderness of my kitchen searching for my alter ego, my other, more talented half, who may or may not have been hiding in the fruit bowl.

I list as many unnatural silences as I can think of—deletions, omissions, abandonments and refutations of subject matter.

Silence is a technology, a pose, a standing position. If maintained over minutes, hours, days, months and years, this 'simple' pose (like standing on a single spot) becomes excruciating. Yet the machinery of silence rolls on.

Speech rises from the fullness of silence. Silence is in, and between, every single word.

To be silent does not mean lacking expression. Silence can be the temple of 'no', can be a crack that happens in the everywhere of life.

Today I try to imagine those who survive the slaughterhouse long enough to hear their own hind legs kicking, like two winds knocking through a wood.

I close my eyes and ask the souls of the dead permission to visit with them.

*O Soul. I wish to see you, hear you. May I cross over? I will prepare a path for myself. I am equipped. My flame will be your flame.*

# Inferno

It is reported that the crisis begins with the mass closure of restaurants, followed by slaughterhouses, due to infection rates among workers. This creates enormous bottlenecks in an inelastic agricultural supply chain. Millions of pigs are all fattened up with nowhere to go.

On a conference call, a co-op of nine farmers agree, though reluctantly, they will start killing piglets. One believes his farm managers will do it by gas or injections. Though he would prefer not to know.

Pigs are fattened inside temperature-controlled buildings. Designed to be slaughter-ready in six months. They can injure themselves if kept alive beyond this time. One pig breaks a leg under the pressure of her own body weight.

A farmer kills pigs that have almost reached the upper weight limit for market, which is 136 kilos. When pigs reach 158 kilos they can no longer be slaughtered commercially. They are too heavy for workers to hoist along the line. They are all dumped in landfill.

With nowhere to ship fully grown pigs, another farmer orders employees to abort the 7,500 piglets that will no longer fit into his units.

To describe the situation of backlogging, Minnesota Pork Producers Association chief executive David Preisler says, *Imagine if in a school the seniors never graduate but the kindergarteners keep coming, and you fill up the gym, the lunchroom and hallways with students.* Only, he is not talking about students.

JBS Foods converts shuttered processing plants into euthanising facilities. Never before have so many farmers had to kill so many animals so quickly.

One day, carts and tanks of carbon dioxide arrive at a farm gate. They are for the sole purpose of gassing sixty-one thousand egg-laying hens. The farmers who usually sell liquid eggs to restaurants tried to shift their market and sell whole eggs to stores. But, as food shortages have set in, so has a shortage in the supply of egg cartons.

Another farmer culls the smallest five per cent of newborn pigs, or 125 piglets per week. He is not the only one. Small bodies can be composted and become fertiliser.

No one will buy livestock because they no longer have the capacity to process them. Unemployment rises. People call for animals to be donated to food banks. But, again, who will slaughter and *process* them? *Processors* are getting sick. Farmers start giving piglets away for free. Still, the price of pork rises 6.6 per cent in supermarkets. Food lines grow. Photographs are posted online of empty, or almost empty, meat aisles.

Conference call. A farmer describes the emotional strain of killing three thousand pigs in a single day.

Chickens are being *depopulated* in the fields. A farmer declines to say how his livestock have been euthanised, just that it was legal and humane. Rural towns are overrun with shame and grief. Farmers are sent hate mail.

In meat plants, workers stand shoulder to shoulder while packing carcasses, in busy locker rooms, walkways and cafeterias. Workers petition for better workplace protections.

One meat giant takes out a full-page ad in *The New York Times*. It says, *The food supply chain is breaking*. It says, *we formed a coronavirus task force, put in place numerous measures to protect our team members across the nation*. Soon it is forced to temporarily close one of its facilities when nine hundred workers test positive for the virus.

In a single state, within a single month, ninety thousand pigs are *eliminated*.

The meat processing workforce in the United States is largely made of immigrants and refugees. Estimates of their hourly wage range from fourteen to sixteen US dollars. I read that if these workers are visa holders, and if they apply for unemployment benefits they may not be eligible for permanent residency under a new rule introduced by the Trump administration.

A hog farmer has a *backlog* of ninety-two thousand pigs waiting for slaughter. They are all dumped in landfill.

A worker at Tyson Foods contracts the virus and dies. Tyson temporarily closes the plant.

More than twenty slaughterhouses close in response to the virus. Trump issues an executive order that they remain open. Meat and poultry executives praise the move. Unions condemn it. The Environmental Working Group says he is marching workers off to death sentences: 4,900 workers have been infected.

Pigs are shot, gassed, administered an anaesthetic overdose or killed with *blunt force trauma*, which means piglets can have their heads slammed against the ground. Other techniques include ventilator shut down, or VSD, with the addition of carbon dioxide. This technique is sometimes referred to as a *modified atmosphere* method of killing. It involves locking a herd of pigs, or a flock of chickens, inside a building and turning off the ventilator systems—the pit pans and the fans. The result is that temperatures rise, gases accumulate, and animals suffocate. Usually this occurs over a period of hours. In extreme cases heat, steam or gas is injected into the building to effectively bake the pigs, or chickens, alive.

VSD is permitted by the American Veterinary Medical Association in *extreme* or *urgent* cases. During the pandemic, some pig producers interpret this guidance as a means to cut costs.

Allegedly, an upper-level manager at a major processing plant directs workers to ignore symptoms of the virus and come to work regardless. He calls the virus a *glorified flu*, said *it's not a big deal* and *everyone would get it*. Managers avoid the plant floor for fear of catching the virus.

Allegedly, a manager at the same processing plant intercepts a sick supervisor on his way to get tested and orders him back to work, saying *we all have symptoms, we have a job to do.*

Another major plant sets up a white tent at the front door for temperature checks. Workers with elevated temperatures are allowed to work. If a worker wants to dodge the tent, they simply enter via the side door.

At one plant, a five-hundred-dollar *thank you* bonus is paid to workers who turn up to every shift for three months in a row, despite rising case numbers.

Cardboard screens are placed around lunch tables and hand sanitiser stations are set up at Smithfield. Workers are given *beard nets* to wear over their faces. Some workers take photos of these beard nets and post them online. These are not the same as surgical or N95 masks.

A man dies alone, in hospital. Age sixty-four. He rarely complained about his gruesome job at his plant, sawing the legs off pig carcasses. I learn that on his final day at the plant, he mopped floors while suffering a fever.

Allegedly, the supervisors of this plant are advised to deny confirmed cases or positive test results. Their workers are told they have a responsibility to ensure Americans don't go hungry.

Another plant offers a five-hundred-dollar *responsibility* bonus to workers who complete their shifts through to the end of the month. This is part of their #ThankAFoodWorker initiative. Workers who had previously decided to leave now decide to stay. Infection numbers are still rising.

A farmer feeds his pigs soyabean hulls, which fill their stomachs but offer negligible nutritional value. He does this to keep them from growing too big. Another alters his feed recipe to make it less appetising. A third turns the heat up in the sheds beyond the comfort zone of the pigs. This slows their eating. It means they are hot and hungry. It increases their discomfort in already uncomfortable factory farm settings.

Reports state that fears of meat shortages *sweep the nation*. A man is interviewed outside a Food Town Supermarket in New York. He has come for chicken but is afraid that he will leave empty handed.

Soon, ten million hens have been smothered, suffocated or drowned with water-based foams that are like firefighting foams. This *humane* method of depopulation is listed as the *preferred* method by the American Veterinary Medical Association. The European organisation Compassion in World Farming suggests farmers use foam that contains nitrogen gas because death comes faster.

To cope with rising infection rates among workers, one factory announces a plan to suspend operations in large parts of its plant on April 11 and will *completely shutter* on April 12 and 13. Workers are called in on all three days of the suspension. The BBC learns that the plant remains running at sixty to sixty-five per cent capacity. That means hundreds of workers go in and out of a building raging with virus.

To perfect their use of ventilator shut down one large pig *grower* performs a test cull with a small number of sows. The pigs are ushered into a shed. The fans are shut off and the heating is turned up to forty-eight degrees Celsius. After five hours none of the pigs have died. The heat is not, yet, fatal.

This story does not end: the company then decides to inject steam into the sheds to accelerate the build-up of heat and humidity. This is how they ultimately perfect their VSD technique. Trial and error.

More than one thousand of one plant's 2,800 workers test positive for the virus. At least five workers at the same plant have already died. Workers stand in cold and damp indoor areas for long stretches of time. The loud noises of the factory floor mean they shout to hear each other, releasing more respiratory droplets.

One farmer seals the cracks in his pig sheds and pipes in carbon dioxide. Another shoots pigs in the head, one by one. It takes him all day to kill the herd. A third has experience gassing newborn piglets, which he is *fairly confident* isn't painful. Though it is, he concedes, disturbing as their legs flail and muscles spasm. Though he notes that there are no vocalisations.

One plant allegedly sends workers from one factory to another, even if they've tested positive. At least one worker vomits on the production line. He is allowed to finish his shift and return to work the next day.

This story does not end: managers direct supervisors to ignore signs of illness in their workers and ensure their *direct reports* turn up for work.

By now, ten million market hogs have been killed and farmers need to dispose of the bodies. Choices include burning, burying or composting. In one state a central facility is opened to run carcasses through woodchippers. This helps speed up the composting process, which would otherwise take sixty days.

This process had been devised in preparation for a possible African swine flu outbreak, but it works just as well for this virus, too. They can chipper the equivalent of two thousand pigs per day.

A video of a VSD cull circulates online. Rumours are the process takes less than an hour and the remaining pigs are shot with bolt guns. What the footage shows is a man with a gun nudging bodies with his foot. In the morning, workers enter the shed to *extinguish* survivors though the number of pigs is so great that usual practices, like checking for a pulse, aren't performed.

Another plant announces it will start euthanising three thousand pigs per day, all of whom are now too big for commercial slaughter, too expensive and space hungry to keep alive.

This small farmer doesn't want to use gas on his herds. He believes he has developed a quicker and more humane method of *depopulation*.

Imagine a mobile unit with a *restrainer*, an electrical stun point and a captive bolt gun. He can slaughter 170 animals every forty-five minutes with a *rotating crew* to avoid mental and physical fatigue. No pig carers are to be involved in the killings, he says. In addition, this farmer donates two pigs to a local animal sanctuary.

A select few plant managers place a *cash buy-in winner-takes-all betting pool* for supervisors and managers who want to bet on how many workers will test positive for the virus. The son of a deceased worker brings a lawsuit against the meat giant after his father dies. The CEO says he is *extremely upset* about the allegations and that the alleged behaviour is *disturbing*, though he is among those personally named in the lawsuit.

This company's slogan: *A cut above the rest.*

# Acknowledgements

This work was researched and written on the unceded lands of the Wurundjeri / Woi wurrung, Boon wurrung and Bunurong people. I pay my respects to their culture and Elders past, present and future. I have lived and worked as an uninvited guest on these lands. I am deeply grateful for these places, the ways they nourish me. I am committed to nourishing them in return.

Thank you, thank you, thank you to Amanda Johnson, Melissa Boyde, Lynn Mowson, Sue Pyke, Laura Jean McKay, Mireille Juchau, Rebecca Giggs, Danielle Celermajer, and the creative intellects of the Multispecies Justice reading group at the Sydney Environment Institute. I cannot think what this work would be without your creative generosity and ethical commitments.

Thank you also to Grace Moore, who provided funds and fellowship to this project by bringing me into the ARC Centre for the History of Emotions as a research associate.

The foundations of this book were laid during my PhD years under the careful guidance of poet-novelist-painter, A. Frances Johnson. Countless conversations, conference papers and two significant research trips have allowed my thinking to expand since that time. I am deeply grateful to have received travel funds from the Felix Myer Creative Writing Travel Scholarship, which supported a writer's residency in France, and the Centre of Excellence for the History of Emotions Travelling Scholarship, which allowed me to take up a fellowship in the Animals in Society Institute at the University of Illinois.

Thank you Bambi, Lurline and Amanda for all your support and especially for not asking when this book would be finished.

To Maria Tumarkin, I have no words to say what it means to have you right there for the good bits, and the worst.

When this book came into the hands of Terri-ann White I knew it would have the most care-full process of development. And it has, it has, it has. I want to express my deepest gratitude to Jo Darbyshire for allowing *The Glorious Decline* to wrap—gloriously!—'round the cover of this book.

This book has been written for farmed animals and in solidarity with those who care for other species as kin.

Finally, and fully, to Laura—so much of what is good in this book is because of you.

Sections of this book have appeared in altered form in *The Animal Studies Journal, The Lifted Brow, Art + Australia* and *Sydney Review of Books.*

# Works Cited

**February 2022**

*Every Twelve Seconds: Industrialized Slaughter and the Politics of Sight*, Timothy Pachirat (New Haven and London: Yale University Press, 2011), p. 129.

'Chapter 5: Food, Fibre, and other Ecosystem Products' in *Inter-Governmental Panel on Climate Change Sixth Assessment Report*, 2022, https://www.ipcc.ch/report/ar6/wg2/

'Introduction' in *The Divine Comedy I: Inferno*, Robin Kirkpatrick (Melbourne: Penguin Group, 2006), p. xliii.

*Wake in Fright*, Kenneth Cook (Melbourne: Michael Joseph, 1961), p. 131.


**To Be Struck**

'What is Contemporary' in *What is an Apparatus? and Other Essays*, Giorgio Agamben, trans. David Kishik and Stefan Pedatella (Stanford: Stanford University Press, 2009).

'The Slaughterer' in *The Seance and Other Stories*, Isaac Bashevis Singer, trans. Mirra Ginsburg (New York: Penguin Books, 1967), p. 30.

'Composition as Explanation', Gertrude Stein, 1926, *Poetry Foundation*, http://www.poetryfoundation.org/learning/essay/238702

*Spectres of Marx*, Jacques Derrida, trans. Peggy Kamuf (New York: Routledge, 1994), p. 11.

## On Immunity

*On the Line: Notes from a Factory*, Joseph Ponthus, trans Stephanie Smee (Carlton: Black Inc. Books, 2019), p. 160.

*The War Against Animals*, Dinesh Wadiwel (Leden: Brill, 2015).

'The Dawn of the "Tryborg"', Jillian Weise, *The New York Times*, 30 November 2016, https://www.nytimes.com/2016/11/30/opinion/the-dawn-of-the-tryborg.html


## August 2021

'Slaughterhouse', Georges Bataille, trans. Annette Michelson, *October*, vol. 36, 1986, *Georges Bataille: Writings on Laughter, Sacrifice, Nietzsche, Un-Knowing*, p. 1929, https://doi.org/10.2307/778539

*Animalia*, Jean-Baptiste del Amo, trans. Frank Wynne (Melbourne: Text Publishing, 2019), p. 257.


## May 2020

'We Fell Asleep in One World and Woke up in Another', Haroon Rashid, *Author Haroon Rashid*, 24 February 2021, https://authorharoonrashid.wordpress.com/2021/02/24/we-fell-asleep-in-one-world-and-woke-up-in-another-author-haroon-rashid/

'The Fairy Tale Virus,' Sabrina Orah Mark, *The Paris Review*, 6 April 2020, https://www.theparisreview.org/blog/2020/04/06/the-fairytale-virus/

'Sarah Manguso on How Lockdown is Re-inventing Slowness', Sarah Manguso, *Frieze*, 4 June 2020, https://www.frieze.com/article/sarah-manguso-how-lockdown-re-inventing-slowness

'Ignorance', Joy Williams, in *Ninety-nine Stories of God* (Portland, Oregon: Tin House Books, 2016), p. 29.

'The Heavy Air', Anne Boyer, *The Yale Review*, 1 December 2020, https://yalereview.org/article/heavy-air

### Back-alley Poets

'Franz Kafka: On the Tenth Anniversary of His Death' in *Walter Banjamin: Selected Writings, Volume 2, 1927–1934,* Walter Benjamin, trans. R. Livingstone (Cambridge & London: The Belknapp Press of Harvard University Press, 1934), p. 810.

'The Gender of Sound', Anne Carson, in *Glass, Irony & God* (New York: New Directions, 1992), pp. 119, 121.

*Unthinking Mastery: Dehumanism and Decolonial Entanglements,* Julietta Singh (Durham: Duke University Press, 2017).

'Murder in the Kitchen', Alice B. Toklas, in *The Alice B. Toklas Cookbook* (London: Michael Joseph, 1954).

'All the Better to Hear You With', Sabrina Orah Mark, *The Paris Review,* 8 September 2020, https://www.theparisreview.org/blog/2020/09/08/all-the-better-to-hear-you-with/

### Manifesto

Manifesto published on *Pink Chicken Project* website, https://pinkchickenproject.com/

### Abandonment

*A Field Guide to Getting Lost,* Rebecca Solnit (Edinburgh and London: Canongate Books, 2017).

### Big Curse Energy

'How to Break a Curse', Danielle Boodoo-Fortuné, *Poetry Foundation,* https://www.poetryfoundation.org/poetrymagazine/poems/150274/how-to-break-a-curse

### Never Destroyed

*Moby Dick,* Herman Melville (Minneapolis: Lerner Publishing Group, 2014), p. 615.

*The Tempest,* William Shakespeare, Scene 5, Act 1, p. 9.

**May 20, 2021**

*Animal to Edible*, Noëlie Vialles, trans. J.A. Underwood (Cambridge: Cambridge University Press, 1994), p. 53. https://www.youtube.com/watch?feature=player_embedded&v=-BluFK9S1I8

**Bright Unbearable**

'In Blue', Eliot Weinberger, in *Oranges and Peanuts for Sale* (New York: New Directions, 2009), p. 147.

*Memorial*, Alice Oswald (London: Faber and Faber House, 2011), p. 1.

**On Red**

*Landfill: Notes on Gull Watching and Trash Picking in the Anthropocene*, Tim Dee (Vermont: Chelsea Green Publishing Company, 2018), p. 18.

**Even Grief is an Immunitary Defence against Animality**

'A Conversation with Koko the Gorilla', Roc Morin, *The Atlantic*, 28 August 2015, https://www.theatlantic.com/technology/archive/2015/08/koko-the-talking-gorilla-sign-language-francine-patterson/402307/

'The Genesis of Mind', Henry Maudsley, *Journal of Mental Science*, vol. 7, no. 40, 1862, pp. 461–494, https://www.cambridge.org/core/journals/journal-of-mental-science/article/abs/genesis-of-mind/92E7FDD7500194E4C43CA871BD31C748

'The Giaour', George Byron, 1813, in 'The Suicidal Animal: Science and the Nature of Self Destruction', Edmund Ramsden and Duncan Wilson, *Past & Present*, vol. 224, no. 1, 2014, p. 206. https://doi.org/10.1093/pastj/gtu015

*A Year of Magical Thinking*, Joan Didion (New York: Knopf, 2005), p. 26.

'A Dissertation on the Origin and Foundation of the Inequality of Mankind', Jean Jacques Rousseau, trans. G.D.H. Cole, pp. 36–37,

https://aub.edu.lb/fas/cvsp/Documents/DiscourseonInequality.
pdf879500092.pdf

*Mortality*, Christopher Hitchens (Sydney: Allen & Unwin, 2021).

'B for Beasts: Do Apes Really ape?', Vinciane Despret, in *What Would Animals Say if We Asked the Right Questions?* (Minnesota: University of Minnesota Press, 2016), pp. 7–8.

*Duck, Death and the Tulip*, Wolf Erlbruch (Wellington: Gecko Press, 2008).

C. Lloyd Morgan quoted in 'The Suicidal Animal: Science and the Nature of Self Destruction', Edmund Ramsden and Duncan Wilson, *Past & Present*, vol. 224, no. 1, 2014, p. 215.


**Hauntings Come …**

'Nevertheless', Joy Williams, in *Ninety-nine Stories of God*, (Portland, Oregon: Tin House Books, 2013).


**Notes on Ghosts**

'A Letter to Sylvia Plath', Ceridwen Dovey, in *Only the Animals* (Melbourne: Penguin Group, 2014), p. 226.

'Hundstag', Dovey, *Only the Animals*, pp. 94–95.

'A Glossary of Haunting', Eve Tuck and C. Ree, in *Handbook of Autoethnography*, eds S.L. Holman Jones, T.E. Adams and C. Ellis (California: Left Coast Press, 2013), p. 649.

'Why Look at Animals?', John Berger, in *About Looking* (London: Bloomsbury Publishing, 1980), p. 15.


**November 2021**

*Notes Made While Falling*, Jenn Ashworth (London: Goldsmiths Press, 2019), pp. 1, 9.

'Word of Mouth: Gossips II', Marina Warner, in *From the Beast to the Blonde: On Fairy Tales and Their Tellers* (London: Chatto & Windus, 1994), p. 28.

*To Write As If Already Dead*, Kate Zambreno (New York: Columbia
University Press, 2021).


**Anatomies of Force**

*Three Guineas*, Virginia Woolf, 1938, https://gutenberg.net.au/
ebooks02/0200931h.html

'A Situation of Meat', Maggie Nelson, in *The Art of Cruelty:
A Reckoning* (New York: W.W. Norton, 2012), p. 175.

*Regarding the Pain of Others*, Susan Sontag (New York: Farrar,
Straus, Giroux, 2003).

'The Illiad, or the Poem of Force', Simone Weil, *Chicago Review*,
vol. 18, no. 2, 1965, http://biblio3.url.edu.gt/SinParedes/08/Weil-
Poem-LM.pdf

*The Sexual Politics of Meat: A Feminist-Vegetarian Critical Theory*,
Carol J. Adams (London: Continuum, 1990).


**Cannibal Cafe**

'Tasteless: Towards a Food-based Approach to Death', Val Plumwood,
*Environmental Values*, vol. 17, no. 3, 2008, p. 324. https://doi.
org/10.3197/096327108X343103

*Francis Bacon: The Logic of Sensation*, Gilles Deleuze, trans.
D. W. Smith (London: Continuum, 1981), p. 17.

'Being Toward Meat: Anthropocentrism, Indistinction, and Veganism',
Matthew Calarco, *Dialectical Anthropology*, vol. 38, 2014, p. 30.

'Too Much Blood', Keith Richards and Mick Jagger, *Undercover*
(Sony/ATV Music Publishing, 1983).

Lyrics transposed from song on Twitter account for Meiwes/Brandes
the Musical, https://twitter.com/meiwesbrandes?lang=en

Melanie Klein and Susan Suleiman in 'Table Matters: Cannibalism
and Oral Greed', Maria Tatar, *Off With Their Heads! Fairy Tales and
the Culture of Childhood* (Princeton: Princeton University Press,
1992), p. 199.

'Table Matters: Cannibalism and Oral Greed', Maria Tatar, in *Off with their Heads! Fairy Tales and the Culture of Childhood* (Princeton: Princeton University Press, 1992), p. 200.


### A Smack of Sorrows

'The Hollow Men', T.S. Eliot, https://allpoetry.com/the-hollow-men

*Weather: A Novel*, Jenny Offill (London: Granta, 2020), p. 41.

'The Transformation of Silence into Language and Action', Audre Lorde, *The Cancer Journals* (New York: Aunt Lute Books, 1980), p. 40.


### Calculus

*Medea*, Euripides, Line 230.

*White Hunger*, Aki Ollikainen, trans. Emily Jeremiah & Fleur Jeremiah (Great Britain: Peirene Press, 2015).


### The Jungle

*Every Twelve Seconds: Industrialized Slaughter and the Politics of Sight*, Timothy Pachirat (New Haven and London: Yale University Press, 2011), p. 54.

*The Jungle*, Upton Sinclair (London: T. Werner Laurie Ltd, 1906), p. 114.


### Dictionary of the Vulgar Tongue

*The Body in Pain: The Making and Unmaking of the World*, Elaine Scarry (New York: Oxford University Press, 1985).

*Pig Tales: A Novel of Lust and Transformation*, Marie Darrieussecq, trans. Linda Coverdale (London: Faber, 1997).

### Snuff

'Not I', Samuel Beckett, 1973, https://www.youtube.com/watch?v=16rSsThMDiU

'The Art of Death: Writing the Final Story', Edwidge Danticat, *Poets & Writers Magazine*, vol. 34, no. 4, 2017, https://link.gale.com/apps/doc/A494442897/LitRC?u=anon~9144800b&sid=googleScholar&xid=462ee9d7

*Every Twelve Seconds: Industrialized Slaughter and the Politics of Sight*, Timothy Pachirat (New Haven and London: Yale University Press, 2011), pp. 257–261.

'Poem on Disappearance', Kimberly Blaeser, *Library of Congress*, 13 July 2020, https://www.loc.gov/item/2020785226/

*Blakwork*, Alison Whittaker (Broome: Magabala Books, 2018), p. 40.

### O Blood as it Pours Down the Drain

'Any Fool can Get into an Ocean ...', Jack Spicer, *Poetry Foundation*, https://www.poetryfoundation.org/poetrymagazine/poems/51258/any-fool-can-get-into-an-ocean-

*Horrorism: Naming Contemporary Violence*, Adriana Cavarero, trans. William McCuaig (New York: Columbia University Press, 2009).

### Joint Labour of Generations

*The Unbearable Lightness of Being*, Milan Kundera, trans. Michael Henry Heim (London: Faber, 1985).

'Inside PMU: Urine Factories and the Menopause Horse Industry,' Sarah Carey, *Horse Talk NZ*, 15 February 2017, https://www.horsetalk.co.nz/2017/02/15/pmu-urine-factories-menopause-horse-industry/

*Timaeus*, Plato, in *Plato: Complete Works*, eds. John M. Cooper and D. S. Hutchinson (Indianapolis: Hackett Publishing Company, 1997), p. 1291.

*Feminism and the Mastery of Nature*, Val Plumwood (London: Routledge, 1993).

## Holiday in Hell

*Every Twelve Seconds: Industrialized Slaughter and the Politics of Sight*, Timothy Pachirat (New Haven and London: Yale University Press, 2011), p. 54.

## Raw/Cooked

'A Lesson in Wisdom in Mad Cows', *We are All Cannibals: And Other Essays*, Claude Lévi-Strauss, trans. Jane Marie Todd (New York: Columbia University Press, 2016), p. 117.

Images referenced are from Nicholas Walton-Healey's *The Disassembly Line*.

*Diary of a Steak*, Deborah Levy (London: Book Works, 1997).

## Inferno

'Coronavirus Crisis Puts Hog Farmers In Uncharted Territory: Killing Their Healthy Livestock', Olivia Solon, *NBC News*, 29 May 2020, https://www.nbcnews.com/news/us-news/coronavirus-crisis-puts-hog-farmers-uncharted-territory-kill-their-healthy-n1216571

'US Food Supply Chain Is Strained as Virus Spreads', Michael Corkery and David Yaffe-Bellany, *The New York Times*, 13 April 2020, https://www.nytimes.com/2020/04/13/business/coronavirus-food-supply.html

'Tyson Suspends Managers at Iowa Plant Who Allegedly Bet on COVID-19 Outbreaks', Ryan J. Foley, *PBS News Hour*, 19 November 2020, https://www.pbs.org/newshour/nation/tyson-suspends-managers-at-iowa-plant-who-allegedly-bet-on-covid-19-outbreaks

A note on form: 'A Thanatography' is a form that works to invoke the unnamed dead. In this way it is related to lament. It is also a prose poem formed as a collage of news items and images laid side-by-side, a strategy that attempts to give life to moments of death and remember the dead in a world that would have us forget.

A note on the fleischgeist: To speak of the *fleischgeist* is to speak of a social and cultural phenomenon wherein humans are haunted by the violence of intensive animal agriculture. What goes on in Concentrated Animal Feeding Operations (CAFOs) and abattoirs ghost people with a deep philosophical disturbance, and so we struggle between acknowledgement of the violence that produces meat and a refusal to look at the conditions in which "food" animals are born, raised and killed.

Fleischgeist is a composite term, a made-up word first offered by Amy Standen and Sasha Wizansky, editors of short-lived magazine meatpaper. Riffing off Hegel's concept of the *zeitgeist*, Standen and Wizansky defined the *fleischgeist* as the dominant spirit of our times, characterised by a 'meat consciousness' fuelled simultaneously by ethical considerations and instrumental logic.

The term also acknowledges that there is a frenzied spirit in the air and in our collective consciousness – a meat mania. With articles that examine meat art, the rise of meat glue, whole-animal butchery challenges, rock star butchers, and the way globalised meat markets identify nations as either 'dark meat' or 'light meat' countries, *meatpaper* stands as testament to the fact that the flesh of other species is used as a metaphor for violence between humans, as art objects, food source, and symbol of national, sexual and gender identities. To read the magazine's twenty issues is to come face-to-face with the world as meatscape.

**About Upswell**

Upswell Publishing was established in
2021 by Terri-ann White as a not-for-profit
press. A perceived gap in the market for
distinctive literary works in fiction, poetry
and narrative non-fiction was the motivation.
In her years as a bookseller, writer and then
publisher, Terri-ann has maintained a watch
on literary books and the way they insinuate
themselves into a cultural space and are
then located within our literary and cultural
inheritance. She is interested in making books
to last: books with the potential to still be
noticed, and noted, after decades and thus
be ripe to influence new literary histories.

## About this typeface

Book designer Becky Chilcott chose Foundry Origin not only as a strong, carefully considered, and dependable typeface, but also to honour her late friend and mentor, type designer Freda Sack, who oversaw the project. Designed by Freda's long-standing colleague, Stuart de Rozario, much like Upswell Publishing, Foundry Origin was created out of the desire to say something new.